Strengthening Environmental Co-operation

with Developing Countries

ORGANISATION FOR ECONOMIC CO-OPERATION AND DEVELOPMENT

Pursuant to article 1 of the Convention signed in Paris on 14th December 1960, and which came into force on 30th September 1961, the Organisation for Economic Co-operation and Development (OECD) shall promote policies designed:

- to achieve the highest sustainable economic growth and employment and a rising standard of living in Member countries, while maintaining financial stability, and thus to contribute to the development of the world economy;
- to contribute to sound economic expansion in Member as well as non-member countries in the process of economic development; and
- to contribute to the expansion of world trade on a multilateral, non-discriminatory basis in accordance with international obligations.

The original Member countries of the OECD are Austria, Belgium, Canada, Denmark, France, the Federal Republic of Germany, Greece, Iceland, Ireland, Italy, Luxembourg, the Netherlands, Norway, Portugal, Spain, Sweden, Switzerland, Turkey, the United Kingdom and the United States. The following countries acceded subsequently through accession at the dates indicated hereafter: Japan (28th April 1964), Finland (28th January 1969), Australia (7th June 1971) and New Zealand (29th May 1973).

The Socialist Federal Republic of Yugoslavia takes part in some of the work of the OECD (agreement of 28th October 1961).

Publié en français sous le titre:

RENFORCEMENT DE LA COOPERATION
EN MATIERE D'ENVIRONNEMENT
AVEC LES PAYS EN DEVELOPPEMENT

* *
*

The OECD Seminar on "Strengthening Environmental Co-operation with Developing Countries", which was hosted by the Government of France and chaired by Mr. Joseph C. Wheeler, Chairman of the OECD Development Assistance Committee, took place at the International Conference Centre in Paris, 4-6 November 1987.

The seminar was attended by over 100 people and included representatives from development assistance and/or environmental agencies in 15 OECD Member countries, representatives of environmental agencies, agencies responsible for finance and planning, universities and research institutes in 21 developing countries (10 African, seven Asian and four Latin American), five international organisations/multilateral development banks and four non-governmental organisations.

The seminar was built around three objectives:

a) To call upon developing country representatives to identify major environmental concerns and the policies, approaches and institutions which are needed to address them;

b) To explore with developing countries how co-operation can be improved in the context of donor-assisted projects and programmes, and to help them address the environmental problems identified;

c) To consider the most appropriate approaches for carrying out environmental assessments on donor-assisted projects and programmes.

The last session of the seminar entitled "Agenda for the Future" was based on the Chairman's conclusions, which were distributed to all the participants and discussed. The final conclusions of the Chairman, amended on the basis of that discussion, are to be found in Part V.

This book is published under the authority of the Secretary-General.

Also available

OECD ENVIRONMENTAL DATA/DONNÉES OCDE SUR L'ENVIRONNEMENT.
COMPENDIUM 1989 (1989) bilingual
(97 89 03 3) ISBN 92-64-03223-1   330 pages   £26.50   US$46.50   FF220.00   DM91.00

ECONOMIC INSTRUMENTS FOR ENVIRONMENTAL PROTECTION (1989)
(97 89 04 1) ISBN 92-64-13251-1· 148 pages   £13.50   US$23.50   FF110.00   DM46.00

"Developing Centre Studies"

FINANCIAL POLICIES AND DEVELOPMENT by Jacques J. Polak (1989)
(41 89 01 1) ISBN 92-64-13187-6   234 pages   £17.00   US$29.50   FF140.00   DM58.00

"Development Centre Seminars"

THE IMPACT OF DEVELOPMENT PROJECTS ON POVERTY. Seminar organised
jointly by the OECD Development Centre and the Inter-American Development Bank
(1989)
(41 88 07 1) ISBN 92-64-13162-0   100 pages   £9.00   US$16.50   FF75.00   DM33.00

DEVELOPMENT CO-OPERATION. Efforts and Policies of the Members of the
Development Assistance Committee. Report by the Chairman of the Development
Assistance Committee. 1988 REPORT (1988)
(43 88 06 1) ISBN 92-64-13174-4   254 pages   £20.00   US$37.50   FF170.00   DM74.00

Prices charged at the OECD Bookshop.

*THE OECD CATALOGUE OF PUBLICATIONS and supplements will be sent free of charge
on request addressed either to OECD Publications Service,
2, rue André-Pascal, 75775 PARIS CEDEX 16, or to the OECD Distributor in your country.*

TABLE OF CONTENTS

6

OVERVIEW OF THE SEMINAR

The Seminar on "Strengthening Environmental Co-operation with Developing Countries" was the first time that officials from both environmental and
development assistance agencies in OECD Member countries met with their
counterparts in developing countries to discuss ways of working together more
closely to better protect and manage the environment in Third World
countries. Prompted by an OECD Council Recommendation of 1986 on "measures
required to facilitate the environmental assessment of development assistance
projects and programmes" [C(86)26] the seminar addressed a number of issues
related to environmental concerns in developing countries and the policy
approaches and institutions needed.

An underlying assumption of the seminar was that developing countries
could benefit from the experience of developed countries in dealing with
environmental problems. In that regard, the importance of integrating
environmental concerns in sectoral development policies (rather than viewing
environmental protection as a clean-up "after the fact") emerged as a key
lesson which the South might learn from the North.

OECD strategies to protect the environment have come a long way since
Western countries first began to become environment-conscious in the 1960s and
1970s. As the then Director of the OECD Environment Directorate, Erik Lykke
pointed out, one of the early guidelines for action in the 1970s was the
"Polluter Pays Principle", whereby the polluter pays for the damage done, and
for prevention and control measures. Today "react and cure" strategies are
practiced in every OECD Member country for investment in, and operation and
maintenance of, pollution abatement facilities, covering measures taken by
both the public and private sectors.

Even while they were taking steps to implement "react and cure"
strategies, Member countries began to realise that it would be more sound,
both economically and environmentally, to anticipate and prevent pollution
before it occurs. And in the past few years, it has become clear that such
"anticipate and prevent" strategies have a better chance of being implemented
if environmental policy is closely integrated not only with national economic
policy but with sectoral policies affecting agriculture, industry, energy and
the like as well.

As far as developing countries are concerned, however, Mr. Lykke
pointed out that it is unlikely that they will be able to take a "react and
cure" approach. Even though many of them have seen widespread environmental
degradation, they lack the money, the skilled manpower and the legal and
administrative structures to deal with such problems.

The only real option available to most Third World countries will almost invariably be "anticipate and prevent" strategies. That is why it is essential for them as well as for their partners in development to determine how they can best draw upon the experience of OECD Member states to protect their threatened environmental resources.

Of all the many grave threats to the environment that have come to the attention of the international community in recent years, such as desertification or the destruction of rainforests, it is not often realised that environmental destruction should be equated with poverty. Yet for the Minister of Natural Protection in Senegal, Cheik A.K. Cissoko, there is little doubt that "poverty is a form of pollution."

"People who are poor," he noted, "are all too easily led to destroy their immediate surroundings just to survive. They cut down forests; their animals exhaust grasslands; they overexploit marginal land; and they migrate to already over-congested towns."

Ironically enough, Cheik Cissoko continued, economic development, too, can have harmful long-term side effects. Because a growing economy uses more raw materials and energy, not to mention synthetic chemical products, this can generate pollution, whose costs are seldom entered into the calculation of production costs.

Faced with problems of such magnitude, members of OECD must show ingenuity and generosity, for they must realise that development and the environment are not two separate challenges, but rather one and the same challenge. "Development cannot proceed if the resource base is constantly being degraded; and the environment cannot be protected if development does not take into account its destructive cost," he said. Nor can these problems be dealt with by fragmented institutions and policies, for they are part and parcel of a complex system of cause and effect. Mr. Cissoko therefore urged participants to reflect on the concept of sustainable development, for it offers a framework that allows the integration of development strategies with environmental policies in their broadest sense.

Some of the causes of environmental degradation in Third World countries were evoked by Michel Aurillac, French Minister for Co-operation, in an appeal for stepped-up environmental co-operation. Over the past 30 years, he noted, the developing countries have had to meet a sharp rise in their food requirements. (Today, the Food and Agriculture Organisation estimates that developing countries need 6 million hectares of additional farm land each year.) As a result, more and more land, including marginal plots of land, has been put to use for food crops and grazing, and this has destroyed the natural vegetation. What is more, wood is usually used to meet energy needs, destroying natural tree cover. All this is taking place in an ecological context of, among other things, fragile soils and harsh climatic conditions.

As the Deputy Secretary General of OECD, Pierre Vinde pointed out, this environmental degradation is all the more serious for the developing countries because their economies are largely dependent on harnessing natural resources. "What are they to do," he asked, "once their means for agricultural self-sufficiency have been destroyed, the soil rendered infertile, the water table polluted, the rivers destocked? It is these countries' very means of survival that are at risk."

"Paradoxically," he continued, "the less industrialised countries are going to be impoverished that much sooner if development plans are not matched by environmental protection measures. One fact is evident: a large number of countries in Africa, Latin America and Asia are already suffering as a result of the mistakes made a few decades ago when man seriously and lastingly jeopardised the ecosystem in exchange for quick profits .... To some extent, we are all responsible, even unwittingly, for what happened in the past. The object now is to avoid insofar as possible making the same mistakes again."

In this connection, developed countries should at all costs avoid basing their dealings with Third World countries exclusively on the profit motive, according to Gilbert Mourre, Deputy Director General of France's Caisse Centrale de Coopération Economique. Nations rich both financially and in technological capacity are duty bound to consider their relations with less developed countries in terms other than those of mere market outlets.

The North has a responsibility to the South in terms of research and development, he noted, to help these countries solve their problems. Obviously, uncontrolled development cannot be allowed to continue for it degrades the environment, nor can development be achieved merely by transferring industrialised country models. "If," Mr. Mourre continued, "developed countries commit themselves to integrating from the outset the best prevention and protection techniques in projects they undertake with their partners in the South, this could constitute a significant step forward."

Environmental Concerns of the Third World

Some of the most pressing environmental issues faced by developing countries today were highlighted by several participants. Among them, the Director of India's Centre for Science and Development, Anil Agarwal, proposed a nine-point action plan for ecodevelopment in India that has potential relevance for other developing countries.

He recommended that immediate measures to safeguard the environmental heritage be followed up as quickly as possible by a development plan designed to ensure the equitable and sustained use of limited resources. For India, for example, it is vital to take innovative approaches to water management and the regeneration of forests and vegetation, for common lands are badly degraded. This kind of approach has the advantage of respecting traditional village structures as well as offering considerable potential for jobs.

The cities, too, constitute a serious drain on natural resources. Projections give India the world's largest urban population by the end of this century, but urban development based on the capitaland resource-intensive pattern of European and North American cities is proving highly inappropriate for India.

Mr. Agarwal concluded with a warning that a national rehabilitation policy must not lead to the impoverishment or displacement of the poorest sectors of the population and called upon the OECD countries to share their wealth of experience in natural resource and land-use management.

Poor land use was just one of the many reasons for encroaching desertification in the Sahel outlined by Dr. Moulaye Diallo, Technical Advisor to the

Prime Minister of the Republic of Mali. The geographical diversity of the region, he explained, means that the scale and nature of environmental problems associated with desertification vary considerably, affecting urban areas in some countries and rural zones in others, while the severity of the crisis ranges from lost opportunities for increased farm output in productive areas to full-scale natural disasters in others.

Overall, however, the Sahel presents a picture of steady deterioration: drought is persistent; existing wood is being destroyed, while oil imports are burdening governments with foreign debt; structural food shortages are aggravated by unsuitable agricultural practices; and socio-economic change is destroying traditional societies.

"Sahelian countries," Mr. Diallo said, "are faced with an impossible situation. The commodities on which their economies are based can no longer contribute to their development. In order to survive they must borrow, with debt leading to further debt. In order to reimburse, they are 'pushing' production, which leads to surpluses and falling commodity prices. Though this may benefit their creditors, it jeopardises their own development."

Though sectoral and general studies have helped to determine action, Mr. Diallo continued, and drought-control strategies have led to a large amount of international aid, the desired results have still not been achieved. He therefore proposed a series of recommendations that would reinforce international co-operation in the region. They encompassed such areas as regional planning efforts, new strategies to halt the advance of the desert, new development programmes backed up by institutional, legislative and communications strategies, as well as energy, water and urban environmental programmes. Support from mulitlateral, bilateral and non-governmental donors will continue to be crucial.

The OECD Club du Sahel is a unique donor group, as Glenn Slocum explained. In co-operation with the Permanent Interstate Committee for Drought Control in the Sahel (also knowm by its French acronym CILSS), the Club du Sahel has generated much of the basic ecological data concerning the region. Major constraints have been identified and include shortcomings in the preparation of development programmes, insufficient popular participation in development, inadequate regional planning structures, ill-defined land tenure, insufficient donor co-ordination and the setting of unrealistic objectives.

The Club, which has already carried out country and sector analyses and assisted in the preparation of national desertification control plans, is now trying to pinpoint the small-scale activities that have proved successful. In this respect, the Club has urged developing countries to organise themselves more effectively and to promote smaller projects, which favour decentralisation and give local populations more say. It has also called on donor countries and organisations to show greater willingness to re-orient their programmes accordingly.

Food security is an overriding concern of not only the Sahelian countries but of almost all the developing countries. Yet some of the developing countries are losing as much as 50 to 60 per cent of their food crops each year to pest attacks. Edward Johnson, of the US Environmental Protection Agency, described the growing reliance of Third World countries on pesticides and the urgent need to share the experience of developed countries with integrated pest management (IPM).→ anti-emw

As Mr. Johnson pointed out, integrated pest management is a complex technology, which farmers must learn to use. Surveillance procedures are complex and laborious, and the development of new varieties has not kept pace with pest resistance to certain pesticides.

The IPM approach therefore requires, among other things, a good deal of research, effective information delivery systems and properly trained extension services. But these are just the things that the developing countries usually lack. Mr. Johnson therefore outlined a series of general principles as a framework for pesticide policy to help formulate regulations on the selection, timing, placement and on the application equipment and techniques, among other things. Essential supports for such policies are training and information programmes.

He also emphasized the importance of regulatory measures to cover product registration, labels and advertising, domestic production and pesticide imports and exports, quality control and residues. Pricing policies on agricultural inputs and commodities, he stated, are another way that governments can influence farmers' choices of pest-control techniques.

There is growing awareness in Africa that over-reliance on pesticides can entail health hazards from direct exposure or via food residues, as well as serious environmental effects and increasing resistance to pesticides. Dr. Eliud O. Omolo, Senior Research Scientist at the International Centre of Insect Physiology and Ecology (ICIPE) in Nairobi, described the work of PESTNET, a regional Pest Management Research and Development Network whose goal is to help African farmers through integrated control of crop and animal pests. A programme of information exchange and dissemination of research results has been set up, and if more financial aid is forthcoming, research will be expanded to cover more types of pests.

Improving North-South Environmental Co-operation

To ensure that aid agencies effectively take environmental problems into consideration, it can be useful for governments to lay down an overall policy framework. Dr. Klaus Erbel of the German Agency for Technical Co-operation described how his government revamped its policy, setting conservation and rehabilitation of natural resources in developing countries as one of the priority aims of economic co-operation projects. In the past few years, the Federal Ministry for Economic Co-operation has developed procedures and checklists, among other things, to facilitate the assessment of environmental impacts in the co-operation projects that are undertaken.

Dr. Erbel stressed that whether the measures are designed to conserve natural resources, to rehabilitate damaged ecosystems or to strengthen recipient country institutions, it is essential that environmental problems be taken into account at the project-planning stage.

In this respect, the role that financial aid institutions can play was pinpointed by Monique Barbut, of France's Caisse Centrale de Coopération Economique, who stated that financial institutions should request that the environmental aspects of a project be included right from the initial design stage. The more thorough the consideration of the environment in prior feasibility studies, the lower the cost. This also applies to the project itself, for experience shows that adding on a system of environmental protection at a later stage is always more costly. In this way, financial agencies can contribute to effective and less costly environmental protection.

Another way to ensure that environmental considerations do not remain purely theoretical but are actually put into practice was outlined by Alio Hamildil of the Niger Ministry of Agriculture and Environment and Olivier Hamel of the French Ministry of Development Co-operation. THey described a new methodology for rural development planning in Niger that is designed not only to strengthen the interface with aid agencies but also to give a genuine say to rural communities.

The model embodies a number of development objectives, the most urgent being community involvement in land-use and resource management and individual action at the farm-plot level. Resolving land-tenure problems often conditions the success of new pilot projects. In the case of Niger, a definition of agro-ecological units based on cantons appears to be the appropriate approach.

To complement this new development model, the authors described a training strategy for field-level and senior officials as well as a network for research that would allow for a longer-term view.

The Role of Environmental Impact Assessment

Having reviewed developing country needs and approaches that can strengthen environmental co-operation, participants turned their attention to the work carried out by OECD to assess the environmental consequences of development assistance.

Frans W.R. Evers, of the Dutch Ministry of Housing, Physical Planning and the Environment, traced the evolution of the concept of sustainable development and explained how Environmental Impact Assessment (EIA) can be a strategic tool for planners and decision-makers alike.

Since 1982, OECD has examined what must be done so that development projects benefit from sufficient environmental assessment input. Work was done to identify, among other things, which types of projects most need environmental assessment and the constraints that hamper developing-country action in this field. Having examined the experience of aid agencies with EIA, a number of procedures and processes was outlined, along with the kind of organisation and resources needed to ensure that environmental assessments can be undertaken satisfactorily. Five key elements were pinpointed: timing, personnel, "scoping", reliable data and monitoring.

Among the examples of how EIA actually works in the field was a description of the Sentani Hydroelectric Development Project in Indonesia by Philip Paradine, Environmental Advisor to the Canadian International Development Agency. On the basis of the assessment, specific recommendations were made that led to significant changes in the project design, well before implementation began, and this without seriously diminishing projected power production. Among the recommendations were the involvement of the local population and adequate compensation for inconviences or losses incurred, as well as periodic monitoring once the project was completed. The Sentani project demonstrated the value of incorporating environmental considerations early on in the planning process, for it allowed changes to be made before the project designs were set. Also very important was the "scoping" of the project, for it focussed attention on the real problems and thereby eliminated unnecessary studies that would have been costly and would have delayed the project.

The Greater Cairo Wastewater project was another example of how useful EIA can be. Mohamed Talaat Abu-Saada, of the Cairo Wastewater Organisation, and Stephen F. Lintner of the US Agency for International Development, explained how the environmental assessment review process enabled the Egyptian and US Governments to evaluate phased implementation strategies for this major project, as well as to select technology, evaluate operation and maintenance questions, and identify complementary projects to ensure sustainable project performance.

Among the important lessons to be learned from this project is the fact that EIA can be used in a cost-effective and timely manner when embarking on major capital projects. It can be used to support programme planning and can increase general awareness of the proposed project, its potential environmental impacts, alternatives and mitigation activities.

Environmental assessment techniques demand skilled manpower, which is often lacking in developing countries. Brian Clark, of the Centre for Environmental Managment and Planning of the University of Aberdeen, presented a framework for training specialists that in large measure reflects the OECD Council Recommendation of 1986.

Environmental assessment and management is still an ill-defined issue, and just how training in these areas should be structured is an open question. In the case of managers from developing countries, the extent and scope of training can be difficult to determine: for example, should training be highly specialised or should it be multi-disciplinary? It will of course vary according to the target group, be it decision-makers, development proponents, officials in charge of environmental assessments or training for trainers.

Mr. Clark presented the advantages and drawbacks of actually training Third World environmental managers in the OECD countries and suggested specific areas where the latter could intervene in accordance with the 1986 Council Recommendation.

A number of participants from developing countries pointed to the need for direct financial assistance from aid agencies for environmental assessment, and in particular for training programmes for government officials in the less developed countries.

During the general discussion, participants called attention to the fact that, given the widely differing characteristics and conditions in developing countries, there is obviously a need for more country-specific research and development. Moreover, there is still a long way to go before environmental protection is recognised as a priority in Third World countries. All too often, environmental advocates are still seen as Don Quiyotes, which underscores the need for greater awareness, especially at the political level.

Environmental values will have to be integrated into the decision-making process, from the level of national leaders right down to individual farmers. A major effort will have to be made to fill a number of specific needs, among which are the following:

-- More and better natural resource economics to adequately evaluate the generally underestimated services provided by natural systems;

-- A better assessment of costs in the short and long term, and of who should bear these costs;

-- Greater use of environmental impact assessment procedures to facilitate the transfer of technologies;

-- Greater attention to alternative development approaches, as opposed to the so-called "modern" models derived from practices in developed countries.

Participants outlined a number of reasons that go to explain why environmental issues are not incorporated into the decision-making process. These include a lack of political will and awareness; inappropriate legislation and institutions; a lack of trained manpower and training programmes; and insufficient tools, data and information.

They then suggested a number of specific actions that could be undertaken jointly by developed and developing countries. These included:

-- A definition of local problems based on indigenous values and traditions rather than on the value systems of the donor countries;

-- More research on local conditions in each Third World country;

-- Small community-level demonstrations to convince the local population of the feasibility and impact of environmental studies;

-- Assistance in developing environmental legislation and procedures, in strengthening institutions and in improving technical expertise;

-- Further exploration of the concept of environmental borrowing in the context of "debt-equity swaps", and relating it to shorter-term economic and financial criteria;

-- Research on the implications of world biomass trade;

-- Better information on the management of renewable resource systems, including planning and pricing;

-- Management of the flow of toxic and hazardous substances or technologies from the industrialised countries to the Third World;

-- More research on low-cost technologies for urban sanitation, transportation and housing construction.

Conclusions

To round up the seminar, the Chaiman of the OECD Development Assistance Committee, <u>Joseph Wheeler</u>, drew the conclusions on each of the three objectives of the seminar.

<u>Objective No. 1</u>: "To identify major environmental concerns and the policies, approaches and institutions needed to address them."

Despite each country's unique characteristics, it can be said that environmental problems in Third World countries are of two kinds: the destruction of the natural resource base and the problems arising from industrialisation and urbanisation.

The need to integrate environmental policies with other sectoral policies has become evident in the OECD countries. This process of integration can come about through the following:

-- "Internalisation", that is, a better reflection of the environmental costs in the prices of resources used;

-- A recognition that sound environmental planning can save the government and the public money;

-- Improved co-ordination among various authorities through more appropriate institutional arrangements;

-- Improved information and analytical tools for decision-making;

-- Increased public participation through awareness-building;

-- The cost advantages of preventing environmental damage;

-- The participation of the private sector in the environmental aspects of development projects.

<u>Objective No. 2</u>: "To explore with developing countries how co-operation can be improved in the context of donor-assisted projects and programmes and to help them address the environmental problems identified."

Participants looked particularly at the areas of pest management, water management and rural development and the specific actions that could be undertaken. To strengthen environmental management in developing countries, technical assistance can help, among other things, in the preparation of environmental legislation and regulations, the organisation of training courses and of long-term technical expertise, and of programmes designed to strengthen public awareness and participation.

Objective No. 3: "To consider the most appropriate approaches for carrying out environmental assessments on donor-assisted projects and programmes."

Recipient governments and aid agencies should initiate assessments early on in the project-design stage and set up joint teams to involve all the interested parties. Prior "scoping" sessions should be conducted to identify the relevant ministries or agencies and the most cost-effective approach, and monitoring should be undertaken during construction and operation.

More generally, it was agreed that the OECD countries should continue to strengthen environmental co-operation with the Third World by studying ways to help developing countries identify and incorporate environmental factors in economic decisions: by supporting the exchange of information among aid institutions and developing countries; and by taking immediate steps to implement the OECD Council Recommendations on Environmental Assessment and Development Assistance.

INTRODUCTION: ENVIRONMENTAL PROTECTION

INTRODUCTION

ENVIRONMENTAL PROTECTION: A "LUXURY" TOO EXPENSIVE TO DO WITHOUT

by
Erik Lykke
Director, OECD Environment Directorate

The high priority accorded to safeguarding and improving environmental quality in OECD Member countries today is reflected in a text adopted by 24 Member Governments of the OECD in May 1987. However, the value attached to environmental concerns should not be considered as a luxury that only the most economically developed countries can afford, for environmental protection also means the preservation of the resource base needed for sustained global economic development. In other words, environmental protection directly concerns developing as well as developed countries.

OECD Member countries are presently trying to promote sustainable economic development by adopting approaches to better integrate economic development and environmental protection, an objective they set during the last meeting of Environment Ministers in June 1985. At that meeting, the Environment Ministers, on behalf of their Governments, declared that they would:

"Ensure that environmental considerations are taken fully into account at an early stage in the development and implementation of economic and other policies in such areas as agriculture, industry, energy and transport."

From "React and Cure" to "Anticipate and Prevent"

The first stage of environmental policies can be characterised as "react and cure". Such policies were developed in the late 1960s and early 1970s when damage done to the environment became apparent in many OECD countries. Concern about what was happening to the air, water, the landscape and other natural resources was new in scope and depth. Large numbers of people began to demand environmental improvements. This led governments to create institutions and to pass laws designed to control pollution, manage resources and improve the amenities of towns and the countryside. Scientific investigations of pollutants and polluting activities were launched to determine the extent and importance of their environmental effects and to establish the basis for regulatory action.

One of the key guiding principles for environmental policies in this period was the "Polluter-Pays Principle", adopted by OECD Member countries in 1972. According to this principle, the polluter should bear the costs of pollution prevention and control measures corresponding to a level decided upon by the public authorities. Its purpose is to stimulate a more rational use of scarce environmental resources.

Experience with "react and cure" strategies, however, quickly demonstrated that they had major drawbacks: they were expensive and they only had their effect after the damage was done. By the mid-70s, it had become clear that it is both environmentally and economically sound to anticipate the possible negative environmental effects of an activity, such as an industrial plant, and to design it in such a way as to prevent pollution before it occurs, thus avoiding the costs involved in cleaning up pollution after the fact.

"Anticipate and prevent" strategies have been implemented through a wide range of instruments, such as technology assessments, environmental impact assessments, land-use planning laws, product regulation and controls of various kinds. It must be stressed, however, that "anticipate and prevent" strategies cannot replace "react and cure" strategies entirely: reactive measures will continue to be necessary, for example, to cope with environmental accidents.

The Next Step: Achieving Integration

The experience gained by OECD Member countries during the last few years has shown that "anticipate and prevent" strategies have a better chance of being implemented if environmental policy is more closely integrated with economic policy and policies in specific sectors such as agriculture, energy, industry, health, land use, tourism and transportation. There is more and more evidence that effective integration of these policies can be promoted by:

-- Recognising more fully the interdependence of the environment and the economy;

-- Identifying complementary objectives and improving co-ordination between relevant authorities;

-- Improving aids for decision-making;

-- Increasing public involvement.

Recognising the Interdependence of the Environment and the Economy

A fundamental reason for the deterioration of our natural and man-made environment is that land and resource owners and users are not receiving the appropriate economic signals about the true costs of utilising these resources. In market-based systems, the absence of markets for clean air or clean water leads to an over-utilisation of those resources because no one pays for using them. In economic systems characterised to a greater or lesser degree by government intervention, this over-utilisation may actually be intensified by government subsidies and controls.

There is, therefore, an urgent need to develop approaches for internalising the environmental costs which are imposed on society by the users of natural and environmental resources. Indeed, this was the objective of the Polluter-Pays Principle, mentioned earlier, which was intended to ensure that the full cost of pollution control and prevention measures was reflected in the costs of the goods produced and marketed.

The Polluter-Pays Principle has been gradually implemented by OECD Member countries. Its application is presently required for investment in, and operation and maintenance of, pollution abatement facilities, and measures undertaken by the public and private sectors within each OECD Member country. There are indications too that countries like the Netherlands, Denmark and Sweden are extending this principle in order to reduce pollution from agriculture.

Other measures are needed to give agriculture, industry and private and public developers of all kinds more accurate signals of environmental needs. In particular, governments should take a closer look at their subsidies, price controls, quantity controls or regulations whenever such interventions may lead to economic inefficiency and environmental losses, e.g. agricultural price-support policies leading to over-production and hence excessive conversion of land for agricultural purposes.

Identifying Complementary Objectives and Improving Co-ordination Among Relevant Authorities

At present, most OECD Member countries have established sectoral ministries or agencies whose policies and programmes are relatively narrowly defined. When environmental concerns came to be recognised in the late 1960s and early 1970s, new specialised ministries or agencies were set up rather than assigning responsibility for environmental matters to existing sectoral ministries or agencies. This reflected the fact that the traditional ministries had failed to take sufficient account of environmental needs and, in many cases, were sceptical of new environmental concerns.

However, the newly created environmental institutions were seldom given the political clout and human and financial resources necessary to carry out their task. In many cases, the most they have been able to do is to act as environmental watchdogs over the older and more established institutions and to obtain, through persuasion, what they could not achieve through power.

The deficiencies of such institutional arrangements in ensuring environmentally sound development are becoming more and more evident. Broader sectoral governmental policies should be adopted which recognise the need to pursue programmes that are environmentally as well as economically sound. Then professional and financial resources need to be committed to ensure that the various ministries take environmental concerns into consideration in their everyday work.

This does not mean that the role of environmental ministries would be less important. Quite the contrary: as the major source of information on environmental protection, they need to advise central economic and other sectoral agencies as the latter integrate environmental concerns into their policies. Moreover, because of their experience and technical expertise, environment ministries or agencies will still have to define and set the various environmental standards and norms which must be adhered to by the sectoral agencies in the planning and implementation of economic activities. All this will require, in turn, a more effective use of existing mechanisms for inter-agency communication and for resolving conflicts.

Improving Aids to Decision-making

The recognition of environmental concerns in sectoral governmental policies can be facilitated by the use of several decision-making techniques. The best known of these are environmental impact assessment (EIA) and quantitative risk assessment. Both aim at providing the decision-maker with information on the possible consequences of a particular course of action before a decision is made whether or not to proceed. They differ, however, in terms of their scope and form. Whereas EIA is concerned with the identification, analysis and evaluation of the consequences of activities for the natural and, sometimes, the social environment, quantitative risk assessment is concerned with the identification, analysis and evaluation in probabilistic terms of the hazards for man and the environment associated with a particular activity.

Nuclear power, pesticides, drilling for oil and gas, dam construction and the siting of hazardous industries are among the areas where both EIA and quantitative risk assessment have been most often brought to bear. Their use has resulted in the adoption of measures designed to reduce potentially harmful environmental impacts and/or risks.

Both, however, have been nearly exclusively applied to specific development projects rather than to programmes and policies. In order to make these tools more relevant, their scope should be broadened.

Increasing Public Information and Involvement

Integrating environmental, economic and other policies can often be facilitated by greater public participation. Active public participation not only serves to increase general public awareness of environmental affairs but can contribute significantly to the elaboration, design and implementation of environmental policies and protection measures.

While public participation may not, in itself, serve to resolve the inherent conflicts between competing national policy objectives, it can serve to clarify the choices to be made and the costs to be borne by alternative policy decisions. In other words, public participation has to be seen as a means for building consensus in areas of environmental controversy and conflict and for improving the quality and acceptability of governmental decisions.

In many instances the public itself often possesses important environmental information. Local environmental groups, hunters, fishermen, farmers and landowners often have detailed knowledge of past and present conditions which can contribute significantly to governmental processes of assessment and review.

Citizens who are asked to accept certain decisions are often more willing to do so if they have been directly involved in their preparation and have been given responsibility for their implementation. The experience of OECD countries shows that public participation, if it is to be productive and successful, has to be a two-way communication process. It is the responsibility of government not only to make information on environmental matters available to its citizens in a timely and open manner, but also to ensure that citizens are able to provide constructive and timely feedback to government.

One key problem for a better integration of environmental policies with other governmental policies relates to the general lack of effective participatory mechanisms for involving the public in the formulation and review of environmental and other policies, as distinct from specific decisions on environment-related projects. Efforts should be made to improve the institutional mechanisms by which citizens can be more directly involved in the choice of policy goals before getting down to specific plans.

The OECD Environment Committee's Work on Integrating Environmental Policies

Leaving aside work on environmental assessment and development assistance which will be discussed elsewhere, we shall quickly review the current work of the OECD Environment Committee which is aimed at integrating environmental policies with agricultural and energy policies, and water resources management with other government policies.

In 1985, a special Group was set up which brought together representatives from both environment ministries or agencies and agriculture ministries in order to develop recommendations on how to promote a better integration of environmental policies with agricultural policies.

The crisis that the agricultural sector is now facing in OECD Member countries, together with a growing concern for the environment, are conditions which can help foster improved integration. The Group has found that opportunities for integration can be divided into current and emerging opportunities. Whereas current opportunities can and should be implemented without delay, emerging opportunities depend upon changes that OECD Member countries are prepared to introduce into present agricultural policies to solve surplus and trade problems.

The work of the Group shows that substantial improvement in integration could be obtained, in the short term, through more appropriate institutional and administrative arrangements and procedures involving a mix of advisory, economic and regulatory instruments. These measures include:

-- Expressing, at ministerial level, a clear commitment to integrate agricultural and environmental policies and ensuring that both types of policy do not have unintended negative effects on one another;

-- Promoting the development of agreed agricultural and environmental policy objectives which are clear, concise, measurable, time specific and, where appropriate, expressed in legislation;

23

-- Encouraging an anticipatory, planned approach to the development of agricultural policies which affect the environment and of environmental policies which affect agriculture;

-- Eliminating contradictory policies, such as the provision of subsidies which encourage farmers to develop wetlands while, at the same time, providing grants to those who agree not to develop them;

-- Promoting public participation in agricultural decision-making to ensure that full account is taken of the environmental consequences of implementing a proposed agricultural policy; and

-- Creating decentralised decision-making and implementation structures which facilitate regional and local integration.

In encouraging farmers to take greater account of environmental objectives and to use environmentally favourable agricultural practices, most countries have a strong preference for using advisory services. It is a voluntary approach which takes account of the economic conditions encountered by the farmer while providing him with information regarding the benefits of adopting environmentally favourable agricultural practices. Experience shows that substantial environmental improvements can be achieved by concentrating those advisory services in problem areas. Voluntary approaches alone, however, are not enough; they must be supplemented by economic and regulatory measures.

For example, subsidies for irrigation water should be phased out to the extent that land and water are used more intensively than would be the case if they were priced at their true economic cost. This would make farmers more aware of the value of such inputs, and would be an incentive to make more efficient use of them.

When the issue is not pollution, the work of the ad hoc Group on Agriculture and Environment has indicated that incentive schemes are proving to be particularly effective in modifying agricultural practices. Under these programmes, farmers volunteer to accept payments in return for a commitment to give careful attention to landscape amenity and nature conservation practices.

Where economic instruments cannot be used, regulations and standards may be needed to prevent environmental problems. In most countries, quality standards for fertilizers, pesticides and food additives have been introduced in order to minimise the risks of immediate and long-term cumulative effects posed by these inputs to the environment. In the case of pesticides, all countries have mandatory registration programmes, though they vary considerably from one country to the next. Efforts to harmonize standards, registration procedures and testing methodologies among OECD Member countries are expected to improve the environment thanks to a more thorough evaluation of agricultural chemicals and an earlier detection of problems.

Emerging Opportunities

Opportunities to introduce environmental considerations into agricultural policies at a more fundamental level are now emerging because of the surplus crisis that OECD Member countries are facing. It is acknowledged that price support schemes have distorted the structure of agriculture, and this is why many countries are now seeking ways to reduce the intensification of agriculture and to take land out of production. The key issue at stake, from an environmental point of view, is the extent to which such changes can be made beneficial for the environment.

Areas in which changes may take place in agricultural policies include the reduction or abolition of production incentives through price support and tariffs, income support, set-aside policies and output quotas. Reduced support for production can be expected to decrease agricultural pollution in areas where the intensity of production is high; however, it will probably be at the source of environmental deterioration in sensitive regions unless environmental protection and maintenance incentives are offered to farmers in these regions. One approach might be to target income support towards regional development and the environment.

An alternative or complementary way of reducing agricultural production consists of taking land out of production through land set-aside schemes. These programmes offer, in the short term, an invaluable opportunity to help reduce agricultural surpluses and to enhance environmental quality, but to be effective, they have to be implemented in a manner which does not lead to soil erosion or lower water quality. Restrictions on production can also take the form of production quotas, which can be double-edged. Generally, they stabilize the negative effects that agriculture has on the environment but, at the same time, they tend to encourage a greater intensity of, and specialisation in production of those farms which remain in the industry. Although further environmental deterioration may be prevented, few environmental gains can be expected from the introduction of quotas.

The Integration of Water Resources Management with Other Government Policies

In the case of water resources management, it has become clear that the decisions taken in various sectors of the economy, such as energy, land management, forestry, agriculture, economic development -- over which water resources managers have little control -- can have significant impacts on water resources.

The objective of a recent project was to identify the institutional arrangements, legal requirements and economic and regulatory mechanisms which lead to an improvement in the integration of water resources policies with natural resources management and other governmental policies. As the findings pointed out, progress will be made towards a better integration of water resources management policies with other natural resources management and other governmental policies if basic resource management principles are implemented. These principles refer both to economic efficiency and to legal, administrative or institutional requirements.

The "User-Pays": A Basic Economic Principle

This principle simply means that the beneficiary or the user of a resource should pay the full social opportunity cost of that use. Under-pricing will lead to an excessive use and misallocation of a resource service.

In addition, this principle can help to reconcile the range of con-flicting interests among various water users. The same body of water, a river for example, is often a resource for a variety of users. Recreationalists boat, fish and swim in it; farmers use it to irrigate their fields; communities for drinking water supply; public utilities to help generate electricity; industry for transporting raw materials and manufactured goods. The demands which these interests place on both the quantity and the quality of the river water can vary considerably and often conflict with each other. If energy policy results in extra costs being borne by recreationalists, then energy customers or agencies are having some of their costs transferred to another sector which, in the interests of efficiency and the environment, should be more properly internalised within the energy sector.

An Appropriate Legal, Administrative and Institutional Framework

The operation of the User-Pays Principle rests upon legal concepts of property rights and on the rights to the use of those resources. These, in turn, determine the type of administrative structure or institutional arrangements that need to be established.

Many of the conflicts arising in the management of natural resources are due to an ambiguity over the existing or presumed rights to the use of those resources. Whatever the mechanism chosen for allocating such rights (market or government), it is essential that a structure of rights to water and other natural resources be unambiguous and as complete as possible.

Many existing institutional arrangements are presently poorly designed to deal with interdependent problems. In view of this and the inevitable problems of overlapping between government agencies, a number of steps can be taken to establish institutional arrangements to support the integration effort and to give it legitimacy and credibility:

-- Objectives for integrating water with other government policies should be articulated which are measurable, time specific and subject to review. Moreover, these objectives should be developed and agreed upon by the agencies responsible for the policy sector being integrated;

-- A related step is to establish the degree of responsibility for each agency which has a legitimate interest in the objectives. The goal here is to establish who has the responsibility to initiate action. Since several agencies often could be given primary responsibility, it frequently is advantageous to think in terms of lead agencies rather than a single lead agency, a lead agency being selected according to the issue or problem under consideration;

26

-- Finally, as situations may probably arise where conflict cannot be resolved by the participating parties, it is essential to designate a third party that has the authority to intervene and arbitrate.

Credibility or legitimacy can be provided in one or more ways. Political commitment by governments provides a clear signal to civil servants that integration is expected. Legislation is also a particularly powerful way to provide credibility and is normally the most stable and enduring. A third means is through a policy or administrative decision, especially when heads of agencies jointly commit themselves in precise terms to an integrated set of policy objectives and activities.

The "Integration of Environmental Policies with Energy Policies" Project

Energy is one of the sectors where better policy integration is essential because its production, distribution and use have a major impact on environmental quality. The OECD emissions inventory shows that in 1980 energy was responsible for over 90 per cent of all major air pollutants. Because of the limited effects of add-on controls or post facto corrective actions, environmental problems cannot be resolved efficiently without integrating the means to solve them at an early stage in energy policies and planning.

Not surprisingly, conflicts between energy and environment still frequently arise not only in coping with existing environmental problems but also more and more in trying to avoid those that may arise from energy facility siting or industrial fuel choices. Past OECD work indicates that economically feasible means exist to avoid several potential conflicts, in particular, through more efficient energy use.

OECD Member countries use different approaches to resolve energy and environmental issues including fuel consumption regulations, economic taxes and incentives, institutional procedures and analytical tools for policy-making (e.g. EIA or risk assessment).

This project, which has just been initiated and which will take the form of country reviews, aims at drawing OECD Member countries' main experiences in three main policy areas:

-- First, the internalisation of environmental costs in energy prices: free markets do not by themselves take into account the cost of using clean air or clean water. Energy costs which do not fully reflect the environmental consequences can lead decision-makers to choose an inappropriate type of energy because it has been under-priced. Therefore, it is the responsibility of governments to make sure that these costs are internalised in energy decisions through appropriate fuel pricing, information, regulations or other instruments. The review will also examine what actions are taken by governments to adjust price distortions arising from subsidies or other fiscal measures and which lead to negative environmental effects;

-- Second, the improvement of the availability of environmental information for use in energy decision-making: major energy decisions take a long time before they are translated into new patterns of energy production and use and into effects on the environment. This time delay can limit possibilities for correcting or mitigating the negative environmental effects from such changes. Consequently, efforts should be made to improve the availability, quality and use of environmental information in any energy decision-making process. The review will address what action is taken by governments to improve the availability, the quality and the use of environmental information in energy decision-making, not only for energy project developments but also for energy policies and programmes;

-- Third, increasing energy efficiency to achieve both environmental and energy benefits: in response to the energy price increases of the 1970s, most OECD governments have achieved marked success in reducing energy use per unit of GDP. Recent studies indicate, however, that much remains to be done. The review will examine in what way environmental considerations could play a greater role in stimulating efforts to increase energy efficiency.

Environment-related Activities of the Development Assistance Committee

For the OECD Development Assistance Committee (DAC), "environment and development" has become an issue of increasing importance. The DAC co-sponsored in 1982 the creation of the "Ad Hoc Group on Environment Assessment and Development Assistance" and subsequently endorsed its findings. The conclusions of this work are reflected in the Recommendations by the OECD Council of 1985 and 1986 on environmental assessment of development assistance projects and programmes. The Council asked the DAC, in co-operation with the Environment Committee, to:

-- Collect further information on the way in which aid agencies of Member countries conduct environmental assessment of their development assistance projects and programmes;

-- Examine how risk assessment can be incorporated in assessing the environmental effects of certain development assistance activities;

-- Prepare a report by the end of 1989 on all measures which will have been taken to implement this Recommendation and on pertinent activities in other international organisations.

The regular DAC reviews of Members' development asssistance policies and practices have increasingly included an assessment of the arrangements made by aid agencies to take environment-relevant aspects into account in their aid programming. Similarly, current DAC work on project appraisal, including criteria and procedures for project selection and design, includes environment as a relevant dimension. The DAC Expert Group on Aid Evaluation covers environmental aspects in development co-operation as an important issue that cuts across major areas of concern.

Potentially Relevant Issues for Developing Countries

Environmental problems in developing countries are already or are becoming much worse than those the industrialised countries now face. At the same time, most developing countries are ill-equipped in terms of money, skilled manpower and legal-administrative structures to deal with them. Developing countries will seldom be able to afford a "react and cure" approach. "Anticipate and prevent" strategies will almost always be the only real option available to them. In adopting such strategies and in moving towards an integration of economic and environmental policies, developing countries can draw upon the experience of OECD Member countries.

Work carried out so far in the OECD Environment Committee has shown that some of the key elements for a better integration of environmental policies with other economic and sectoral policies are:

-- Ensuring that environmental costs are reflected in the price of such resources as air and water (by a better application of the Polluter-Pays Principle and the User-Pays Principle) as well as eliminating the various governmental interventions which lead to under-priced environmental and natural resources;

-- Identifying complementary objectives between environmental policies and other economic and sectoral policies and improving co-ordination between relevant authorities through more appropriate institutional arrangements;

-- Improving the analytical tools likely to facilitate decision-making, together with efforts to improve the availability and the quality of environmental information to be used during the decision-making process;

-- Increasing public information and involvement.

29

PART I

SOME ENVIRONMENTAL CONCERNS OF THE THIRD WORLD

ECODEVELOPMENT:

A NINE-POINT ACTION PLAN FOR INDIA

by
Anil Agarwal
Director, Centre for Science and Development,
New Delhi, India

In most developing countries, a national programme to protect the environment must inevitably become an ecodevelopment programme. Merely conserving the environment is a far simpler problem than ensuring its rational use. In a country like India, where both population and development aspirations are growing rapidly, the biggest challenge is to learn how to exploit the environment at higher levels of productivity than now prevail, but in such a manner that this increased productivity can be sustainable. Technically, it should be possible to do so because the declining productivity of most ecosystems today is the result of environmental mismanagement.

The following nine-point programme of action for India, or at least part of it, may well be applicable to other developing countries.

Point 1:   Protecting Natural Resources from Excessive Development Pressures

The first point on an environmentalist's agenda must be to safeguard whatever is left of air, water, land and forest resources. This means that legislation like the Forest Conservation Act, the Air Pollution Control Act, the Water Pollution Control Act and the new Environment Protection Act must all be strictly implemented regardless of political and business pressures.

While these acts could be strengthened, it is more important that they be strictly implemented. This is the major responsibility of government agencies. But where government agencies fail to take up the cudgels on behalf of the environment, individual citizens and citizens' groups must appeal to public opinion and use the legal framework available to them. (In Kerala, for instance, the Kerala Sastra Sahitya Parishad organises street plays to heighten environmental consciousness and played a key role in preventing the damming of the Silent Valley.)

Point 2: Equity and Sustainability in Environmentally-sound Planning

Protection represents only a small part of India's environmental problem. The greater challenge is to improve the standard of living of the vast majority of the people by learning how to use natural resources while generating a higher level of productivity.

With environmental awareness growing, three objectives must now guide the country's development process: growth, equity, and sustainability.

First, the new objective of sustainability demands, in a country with numerous and diverse ecosystems, that planning be decentralised, proceeding differently for each ecosystem. India cannot plan for the high Himalayas as it plans for the Sunderbans, the Deccan Plateau or the Northeast. Development cannot function like a juggernaut.

Second, sustainability demands that prior to the planning and investment stage for a given area, there must be an idea of its 'sustainable economic future' using known technology. This will call, most of all, for good land-use planning for every agro-ecological zone of India.

Take, for example, the district in the Deccan Plateau in which nearly 650 000 hectares are being farmed out of a total surface area of 750 000 hectares. The district today looks like a desert, and its crop yields are abysmally low. There is no overview of the multitude of public works programmes in the district or of whether the net effect will be ecological enrichment or ecological destruction. No one knows how much tree or grass cover is needed to support agriculture in the area. Should that area be used for agriculture to this extent, or in growing grasses and animal care, or in forestry; or an appropriate, more productive and sustainable mixture of all three? Only once the "ecovision" for an area is clear can plans be formulated to turn that vision into reality.

The concept of such 'sustainable economic futures' needs to be developed for every agro-ecological zone of India. Planning must begin from the bottom and must be based on principles of sustainable and equitable development for every ecological region.

Point 3: The Employment Potential of Programmes for Ecological Regeneration

Fortunately for India, ecological regeneration, whether it be soil or water conservation programmes, or afforestation, is a very labour-intensive exercise. Once a decentralised ecodevelopment plan is available for a particular ecological region, massive employment can be created through such activities. If legal employment guarantees can be tied to programmes of ecological regeneration, a massive bid can be made to meet people's needs on a sustainable basis.

Maharashtra, for example, is the only state which guarantees a job as a legal right under its Employment Guarantee Act. Today millions are employed in drought relief projects, most of which are ad hoc in nature. But drought relief projects should provide more than relief _during_ drought; they should provide relief _from_ drought.

Point 4: Emphasizing Equitable and Holistic Village Ecosystems

Though it may sound technical, holistic village ecosystems correspond to what Gandhi once advocated: that India's villages become self-reliant in food, fodder, fuel, biomass-based building materials and raw materials for artisans.

Every village in India is what scientists call in technical jargon an integrated 'agrosylvopastoral entity'; that is, every village has not just cultivated fields, but also grazing lands and forests, groves and trees. These three systems interact with the water system to create a highly complex life-supporting ecosystem. Our objective should be to develop the productivity of all three major land systems, without destroying the productivity of one at the expense of the other. Holistic enrichment of village ecosystems means that India must plan at the same time for the private and common (including government) lands in the villages in such a way that ecological security is maintained and all the diverse fuel, food, fodder and other basic biomass needs of the people are met on a priority basis.

Any biomass production in a village ought to observe the following order of priority: ecological security; basic survival needs; and income generation and export. Today the order of priorities is completely reversed. Farmers plant trees or crops first for export, then to satisfy basic needs and never for ecological security.

How the holistic enrichment of a village ecosystem can be achieved is illustrated by the village of Sukhomajri, situated in the hills near Chandigarh. Ten years ago the area surrounding the village was badly degraded, and in one downpour farmers would lose half their fields. A soil conservationist tried to persuade the villagers not to graze their animals in the watershed area. The villagers argued that there was no other place for their animals to go.

It was finally realised that farmers would do something only if they benefitted from it. Every farmer is hungry for water, and so the project staff built a small earthen dam. The first monsoon came and water collected behind it. But the monsoon and the rain-fed crops of Sukhomajri began to die. The villagers came running to the soil conservationists for water, who offered them a deal: "You'll get water if you stop grazing your animals in the catchment area." The villagers agreed. Since then, crop production in the village has gone up from three quintals to some 17 quintals a year, and now instead of one unreliable crop, the irrigation assures three crops yearly. From a food-importing village, Sukhomajri has turned into a food-exporting village.

But to everyone's dismay, the soil conservationists found that some villagers were still taking their animals into the catchment area. These were landless people who claimed they had not benefitted at all from the dam. They in fact pointed up a grave failure of India's entire water management system, which reflects the inequities in land distribution. It was decided for the first time in India that all households, landed or landless, would get an equal amount of water. Those without land either sell the water to landowners or enter into a sharecropping arrangement, exchanging water rights for a share of another villager's crops.

35

Now everyone in the village protects the watershed, and trees and grasses are growing again. This has increased the availability of fodder greatly and Sukhomajri now sells some 300 000 rupees' worth of milk every year. Most people have got rid of their goats and now own fat, high-yield buffaloes.

With trees available once again, the fuel problem for cooking has eased. Simultaneously, the villagers are planting bhabhar grass in the watershed, which is good for making rope as well as baan, a rope-like material for making simple beds. It can also be sold to papermills. Sukhomajri is now too rich to make its living by rope-making but in a nearby village, several families now earn 400 to 500 rupees a month making rope from the bhabnar grass which they can obtain from their neighbours.

Thus, improvement of the village ecosystem has not only greatly increased food production, improved animal care and milk production, but has reduced fuel problems and revived traditional handicrafts. Environmental improvement has dramatically changed the economic future of Sukhomajri in a short six-year period, demonstrating that such environmental improvements are not a brake on economic development, as is often believed.

A holistic approach to village ecosystems, as in Sukhomajri, points up the importance of two things in particular:

i) A new approach to water management; and,

ii) Massive regeneration of trees and grasses on common lands.

Point 5: A New Approach to Water Management

To improve and enrich village ecosystems, India will have to take a new approach to water management, moving away from the all too prevalent civil engineers' emphasis on big dams and canals towards a multi-disciplinary understanding of hydrology, ecology and political economy.

The fundamental principle of water management should be to catch every drop of rain where it falls. If it falls on a slope, there should be trees to hold it. If it falls in fields, there should be a small dam or reservoir. If it falls in a village, there should be a village pond or tank. Even today India uses only 10 per cent of all the rain that falls on the land.

The second fundamental principle must be equity. Water is a scarce commodity and will become scarcer with time. It must be used with maximum efficiency for the maximum public good. Sugar cane, for instance -- a crop which uses water intensively -- cannot be permitted in semi-arid areas.

Point 6: Massive Regeneration of Trees and Grasses on Common Lands

India has about 100 million hectares of common lands. The maximum environmental destruction today occurs on just such lands, where there is no individual ownership, and most are in a highly degraded state. Centuries ago, the community was conscious of the need to care for common lands but British policies of forest nationalisation destroyed all that.

Multipurpose trees which can hold the soil and water and yield fuel, fodder and food, should be planted on these lands, rather than eucalyptus, pine or teak planted for the sole pupose of meeting urban-industrial demands. But the grazing pressure in India is intense: with only one-fortieth of the world's land area, India supports over half the world's buffaloes and a seventh of its cattle and goats. As multipurpose trees are browsable, they must be protected just like wheat and rice.

If multipurpose indigenous trees are to grow on the commons, social and political conditions must be created so that people can once again develop a sense of responsibility for these trees.

There are two ways in which this special relationship between the people and such lands can be created:

i)   By further privatising the commons (through contracts for the care of individual trees or by leasing out large tracts to private industries); or

ii) By enriching them ecologically while retaining them as commons.

On Gossava Island in the Sunderbans, for example, some 100 to 200 trees along the roadsides have been allotted to a specific landless family, which now cares for them. These roadside plantations are doing well. In Rajasthan, Udaipur, the poorest tribal families were identified and given two hectares of barren forest land every year provided they planted and cared for the trees. The trees are growing well, and the tribals have already benefitted greatly by selling the grass. When the trees finally mature they will benefit from that too.

The problem with this first approach is that it will generally lead to single-purpose trees and monocultures. And as commons constitute the survival base of the poor, this will impose even greater hardship on landless and marginal farmers, who will have to subsist on still smaller commons.

Instead of privatising commons and turning them into small parcels of land that are handed over to individual families, an even better approach is to organise communities to plant and protect trees together on common lands. The second approach is very difficult and has been tried only in a very few places. But the results have been outstanding. In the Chamoli districts, ecodevelopment camps have been organised in which people, particularly women, come to discuss their daily problems. They have taken up tree planting enthusiastically, and the survival rates of the seedlings are well over 90 per cent. Grasses grow quickly on this land.

The women plant trees as a community effort and share equally. They have devised a very simple and equitable distribution system: once a month, the head of the village council designates the day when someone from a given household can take as much grass as he or she can. Invariably, however, the sarpanch, a local political figure, creates difficuties. Threatened politically, he tends to obstruct the formation and activities of popular participatory movements among the poor.

The success of community afforestation efforts on common lands depends on three principles:

i) Control by the interested group over the piece of commons to be ecologically enriched;

ii) Unity within the group or, if the community is stratified because of village politics, within the group that takes up the ecological enrichment; and

iii) Equity in the sharing of the produce from the commons, without which unity is not possible.

Bureaucrats alone are unlikely to do the job. The role of the political system in fostering such social organisation is paramount.

Point 7: Rational Resource Use and Urbanisation

Between now and the year 2000, India's urban population will grow more quickly than the rural population, and by the end of the century, India will have the world's largest urban population -- well over 300 million people.

Current patterns of urban development are based on imported models from Europe and North America and are highly capitaland natural resource-intensive. Capital-intensity divides the urban population into the urban rich who can afford the fruits of science and technology and the urban poor, who barely eke out a living.

Intensive resource use ensures that as cities grow, the rural hinterland from which the resources are extracted will be either environmentally destroyed or transformed. For example, original forests are first destroyed and if regenerated at all, are replanted with eucalyptus, pine or teak. The rural people are deprived of their survival base, forcing them to become "ecological refugees."

Managing the world's largest urban population will be an immense challenge, and the answer lies in technology choices built around the principles of equity and resource-rationality. Take, for example, the housing situation, which is becoming worse in most Indian cities.

The low-cost housing of brick and cement that urban development authorities build are both climatically bad and psychologically demeaning. Moreover, brick and cement are both produced in a way that is environmentally destructive. Meanwhile, beautiful mud buildings exist all over India. Surely it is possible to design mud buildings for cities that are more environmentally sound, equitous and climatically better? It is really a question of social values.

Let us also take a look at water supply, which is becoming a major problem in all Indian cities. Jodhpur gets only a few inches of rainfall, but when the rains come, there is flooding. This annual flood used to be caught by a wide network of canals and collected in tanks and ponds, thus enabling this desert city to survive for centuries. But instead of preserving this time-tested system, modern Jodhpur is destroying it. Waste is now dumped into the canals, and tanks are filled with stinking slime. Meanwhile, the entire catchment area for these tanks is being ecologically devastated through extensive quarrying, which totally disrupts the water flows. Today Jodhpur is no longer self-reliant for its water. It depends for its supply on a distant dam and a 100-kilometre long canal. Urban planning must take into account the health of the environment which sustains the towns and provides city dwellers with the natural resources necessary for their survival.

India must also find socially appropriate technologies for its urban settlements. Take sewer technology, which developed in Europe during the colonial period. In India, however, it is too capital-intensive for it to be accessible. Nearly one third of India's urban dwellers have no access to toilets, while another third has only bucket-type latrines. The last has sewers, but even this is largely serviced by community toilets. In most Indian towns sewers are under-utilised and choked.

Surely composting toilets can be developed that would put a halt to the pollution of the Ganges and other rivers. Cities can, in fact, become major sources of crop nutrients (in the form of sewage sludge) and not just a drain on Indian croplands.

The same applies to urban transport. India does not have a policy to provide reasonable access to transport for all. If it did, bicycles would be ubiquitous. Not only are they healthy and pollution-free, but they are also a more equitable mode of transport.

All this does not mean that India should think only in terms of simple technologies, often labelled as second-grade. In any city buses are needed for long-distance transport, but there is no reason why India's public buses should be so dilapidated. If Indians live in mud houses, that does not prevent them from having access to the finest satellite communications systems. Why then cannot principles of biotechnology and genetic engineering be used to make composting toilets a success?

The basic trouble is that for public services we tend to use the least sophisticated technology while the technologies used in private life are the most modern. Private houses must be made of modern bricks and cement but public telephones never work. Private cars are ever more sophisticated but public buses and transport systems are ever more disorganised and dilapidated. The reverse should be true.

Point 8: Environmental Education

Ultimately, environmental education is essential. The people must become environmentally conscious. Educated urban dwellers are today so alientated from their environment that they have become "resource illiterate". They are the biggest consumers of natural resources and can do enormous environmental harm, depriving the poor of their own environmental needs. The poor, in turn, want to adopt the consumption patterns of the town dwellers. Conservation-oriented education is needed. But rather than dealing with environment as a separate subject, its lessons must be integrated into the teaching of all other subjects, from history and geography to engineering and medicine.

Most important of all, modern Indians must be taught to appreciate and apply the traditional practices that fostered conservation. These practices were often built around religious precepts. The challenge lies not in decrying them as superstition but in maintaining and lifting them onto a secular plane.

Even before there was any population explosion or development explosion, the Indian people were conservation-oriented. The story goes that some 200 years ago, over 350 Bishnois died in Rajasthan protecting Khejari trees from being cut by the king's men. This place is an important pilgrimage for environmentalists, and to this day Khejari trees are never cut in Rajasthan even when they grow in the fields.

Point 9: A National Rehabilitation Policy

If dams are to be built and if industries and cities are to expand, India must develop a national rehabilitation policy soon. More and more people are being displaced by cities, by industries, by mines and especially by large dams. This proposed policy must make displacement and resettlement exercises honest and open, respectful of people's economic and cultural needs and their demand for land. Development must not impoverish anyone.

The hardest hit are the tribal people, who depend on their lands and forests for their survival. For instance, the tribal people near Ranchi have been opposing dam construction for nearly 20 years now. They are very poor people and when asked why they will not leave, they say they would not know what to do with the compensation money. "Our land may be small but it is our anchor. We can somehow eke out an existence". Because of this strong attachment to the land and forests, it is vital that India develop a national rehabilitation policy as soon as possible.

Conclusions

This nine-point action plan is neither comprehensive nor definitive and more thought on ecodevelopment planning methodologies is urgently needed. But it could provide a workable framework for planning equitable and sustainable development. It is clear that primary action for such plans must occur within the country itself and with its rural and urban communities. However, OECD countries can help greatly.

First, it is important to create an awareness within the OECD countries of the adverse land-use changes and pollution that their own consumption is creating in the developing world. Current trading patterns and debt burdens can easily lead to extensive and rapid land-use changes, causing increased mismanagement of the country's natural resource base. UNEP has calculated that a 1 per cent increase in interest rates adds approximately $5 billion to the debt burden of developing countries. To increase its export earnings (not profits) by $ 1 billion in 1981, South America as a whole would have had to increase its banana exports three-fold, Ecuador five-fold and Colombia nine-fold, while leading cotton exporters would have had to double and triple their cotton exports. This would have meant bringing millions of additional hectares into production to grow these export crops.

Agencies in OECD countries should give priority to finding solutions to these problems. In India, the need to earn foreign exchange has led to several environmentally damaging activities. Tea gardens, for example, situated on high Himalayan slopes have expanded at the expense of tropical forests and have reduced the land available for traditional rotation of crops. The export of tea has further required plywood for tea chests, which has wreaked havoc on forests in northeast India.

Second, OECD countries could help developing countries by providing information about the following, among others:

i) Land-use management systems, pollution control management systems, environmental impact assessment programmes, and the like which are used in their own countries;

ii) Possible adverse impacts of toxic products and substances found in trade and commerce; and

iii) Available and emerging appropriate and environmentally benign technologies which could be used to solve a variety of urban and rural problems.

It is not enough to transfer information to developing countries. The Golden Rule should be to transfer information as part of a programme which aims at the development of indigenous skills. The information transferred can be used in a local perspective and appropriate management systems developed accordingly. It may be necessary, as part of this effort, to develop local research institutions which can integrate ecological considerations into economic and technological planning and develop suitable technologies. Where necessary, OECD countries should arrange for transfer of environmentally benign technologies on easier terms.

DROUGHT AND DESERTIFICATION:

A. THE SAHEL: A MULTI-PRONGED PROBLEM

by
Dr. Moulaye Diallo
Technical Counsellor to the Prime Minister, Republic of Mali

Drought and desertification have so scarred the Sahelian countries as to upstage its relatively minor problems of chemical and industrial pollution. Analysing the environment of West Africa's Sahel, a region that covers 5.3 million square kilometres with an estimated total population of 36 million, starts with what has been called "problem areas". The diversity of conditions, uneven penetration of the modern economy, unequal distribution of equipment and other factors go to create various types of problem areas in each Sahel country. These are essentially: natural disaster areas; demographically threatened areas; urban areas; backward areas; and areas of potentially high agricultural output.

The pattern can be more detailed but what matters here is that each situation requires a solution of its own in terms of ecology, public health, education and the economy.

Natural disaster areas. The northern part of these areas is characterised by arid, harsh climatic conditions and also by desertification, rendering it particularly vulnerable and sensitive to water shortage and making it one of Africa's most severely drought-affected regions. The human population follows one of two patterns:

-- Nomadic livestock farmers who are scattered and who return regularly or intermittently to given locations;

-- Concentrations of pastoral farmers around permanent water sources (oases, wells, ponds, lakes, etc.).

Those populations live in and with the desert.

In the southern part (Sahelian, South Sahelien and North Sudan) drought can take several forms; either rain is irregular, below normal levels or absent altogether.

Within these areas, there are basically three types of production systems: a pastoral system in which over half the producers' gross earnings

42

(value of own consumption plus marketed consumption) or over 20 per cent of domestic food calories derive directly from livestock and related activities; an agro-pastoral system in which users draw only between 10 to 50 per cent of their gross earnings from cattle, i.e. at least half from agriculture; an agricultural system in which livestock accounts for less than 10 per cent of earnings.

Demographically threatened areas. Here one finds wide stretches of desert or semi-desert areas; the climate is so unfavourable that settlers tend to move southward towards the Sudan-Sahelien and Sudan areas, which receive a little more water. Malnutrition, morbidity and mortality rates are dangerously high. Other regions prove difficult to live in because of remoteness and difficulties in procuring supplies of water, factors of production and other essentials. Lastly, there are regions that are "under attack" from endemic infectious and/or parasite diseases like onchocerciasis (river blindness).

Urban areas. Urban areas exhibit high and continuous population growth (5 to 6 per cent per year, as against a national average of 2 to 3 per cent), posing new problems for Sahel governments in supply, transport, and distribution of foodstuffs in the towns, rising unemployment, poor housing and public health conditions and delinquency. Urban centres have directly exacerbated pressures on the resources of the surrounding countryside and indirectly affected the remoter areas which provide them with food, fuel, manpower and construction materials.

Backward areas. They are characterised by large-scale migration of the young to towns and to more prosperous areas.

Areas of potentially high agricultural output. These are generally well watered areas with high or rapidly growing populations; vegetation is still relatively plentiful. Deforestation to satisfy the needs of other areas for firewood, together with bush fires, land clearance and migrant arrivals all help to create problems.

In addition to difficulties in procuring factors of production and low literacy rates, these areas all have common problem areas that can be broken down as follows:

-- Climate (direct and indirect deterioration);

-- Ecology (desertification, over-exploitation of ligneous resources);

-- Food (famine, malnutrition);

-- Energy (difficulties in procuring firewood, the main energy source);

-- Demographic (uncontrolled growth rate exceeding that in production, breakdown of traditional societies).

Included in this category are those areas which, though climatically arid, have land and water resources which, once deployed, could provide potentially high yields (Senegal, Niger Valley).

Deteriorating Environmental and Living Conditions

Land degradation and desertification are gathering pace. (Land degradation is the impoverishment and deterioration of the environment's biological production potential, i.e. soil, water, fauna and flora; while desertification is a physical phenomenon -- masses of sand are moved by the wind, which submerges everything in its path, from soil, plant cover and watering holes to buildings, etc. and leaving behind totally sterile areas. This is occurring on the edges of the Sahara desert, which is therefore advancing.) Meanwhile, prolonged drought is reducing the availability of surface and groundwater in terms of volume and duration, hence reducing biological productivity and the resistance of plants, animals and men.

One cause of this state of affairs is over-farming, the result of uncontrolled population growth which increases pressure on the land, and also of the monetisation of the economy. They have incited farmers to produce more than their basic needs.

Unsuitable agricultural and production systems for meeting the requirements of the populations and the impact of the continuing drought largely explain the structural food shortage. Meanwhile, the systematic introduction of crops such as cotton and groundnut leaves the soil bare and loose after cropping. The soil is then easily eroded by the wind and by the sudden rainfalls during the winter. Moreover, harvesting is being done carelessly, without soil-protection measures.

Ironically, because of improved veterinary practices, the number of livestock has grown. More waterholes have been dug, but without controlling levels of use and without adjusting pasture runs.

Deforestation is another major problem. Forests are cleared in order to plant crops, and to meet fuel requirements, so much so that the areas surrounding towns and villages have lost far too many trees. The problem is compounded by shepherds who lop tree branches in order to feed their herds and by men and women eager to sell the leaves and fruit in bulk for meagre profits.

Deforestation is spreading through the Sahelian countries like a disease. It exposes the land to the direct and immediate impact of sun, water and wind.

Owing to their scale and repetition each year, bush fires are one of the main causes of ecosystem degradation. There are three main ecological effects:

-- On vegetation: the natural flora with a balanced composition shifts towards a pyrophilic flora since fire-resistant species no longer face competition from plants that are sensitive to fire and therefore spread. At the economic level, the following adverse effects may be noted: destruction of pastureland, wild produce and wood products;

-- On soil: the many different effects range from direct destruction of organic matter (litter and humus) to indirect destruction through the incineration of fuel plants;

-- On the microclimate: when the plant cover and surface organic
   matter is destroyed, the soil is exposed to the direct effect of the
   weather (sunshine, heat, rainfall, evaporation). Relentless
   sunshine on bare soil harms the microflora and pedofauna.
   Desertification has resulted in the recent development in the
   Southern Sahara of typical desert formations: advancing sand dunes,
   regs, and areas devoid of any soil, vegetation and water
   whatsoever. The desertification area does not show a continuous
   front line; it consists of different sized ring shapes separated by
   Sahelian "corridors" affected by drought but not yet turned into
   desert.

The destructuring of traditional societies as a result of the socio-
economic and socio-political changes over the past century has accelerated
during the recent droughts. These societies are no longer able to control the
management of their land and ecological heritage.

The regional balance and the development basis and systems are
jeopardised but it is by no means certain that once the drought ends, it will
be possible to re-establish the previous state of affairs. There are several
reasons:

-- First, the areas under dry cropping and many areas under irrigated
   cropping with partial water control are shrinking. Herds are moving
   from the North to the South to the detriment of regions primarily
   used as pastureland. The use of fertilizers is limited, hence there
   is little room to improve soil productivity and few opportunities
   for reducing the area under crop per capita.

-- Sales of agricultural equipment have dropped and the use of
   agricultural equipment is declining. Lastly, the population has
   been drifting to the main urban centres and to the South (part of
   the displaced populations are obliged to abandon arable farming and
   livestock production without being able to find other production
   activities).

A Structural Food Shortage

As cereal production is practically stagnating and vulnerable to
drought, food dependency is on the rise. In 1961-1985 production rose by only
0.5 per cent on average whereas the average population growth was 2.5 per cent
per year and urbanisation 6 per cent per year.

Table 1 shows trends in imported food aid to the Sahelian countries
from 1970-71 to 1985-86. Whereas the optimum daily ration is 270 kg of millet
equivalent per year per person and 21 grammes of animal protein per day, most
Sahelian countries can only supply on average 150 to 170 kg millet and 15
g/day of protein per person. The basic shortage is therefore on average 100
kg millet/year and 5 g animal protein/day.

Table 1 CEREAL PRODUCTION IN THE SAHEL

in thousand tonnes

	Burkina P* I**		Mali P I		Mauritanie P I		Niger P I		Senegal P I		Tchad P I	
1961-1965	932	10	1028	10	-	6	1217	4	607	234	600	4
1981-1985	1183	71	1043	170	52	212	1501	96	799	557	403	53

	TOTAL SAHEL P	I
1961-1965	4384	289
1981-1985	4980	1303

*P =total cereal production in 1961-65 and 1981-85
**I =total cereal imports in 1961-65 and 1981-85

Rising imports (including commercial imports and food aid) at an average growth rate of 8 per cent have modified consumption patterns. For instance, local cereals produced during the rainy season (millet, sorghum, maize) account for less than 10 per cent of the cereal ration at Dakar and rice constitutes about half the daily ration at Bamako. Rice and bread consumption is also spreading in rural areas.

Requirements for water of acceptable quality are also far from met. At best half the water supplies are satisfactory. In all the Sahelian countries, modern wells, bore holes and pumps are in short supply despite the considerable efforts made in recent years.

An Increasingly Acute Energy Crisis

In the Sahelian countries fuel wood is central to energy balances (60 to 90 per cent) and in many towns wood purchases usually account for over 25 per cent of the household budget. Meanwhile, oil consumption, estimated at over 300 000 tonnes, is growing rapidly and oil bills are jeopardising development work. Apart from perhaps Chad, all these countries lack fossil fuel resources and their hydropower potential remains mostly untapped.

As an indicator of development and the industrialisation level, the annual consumption of energy in coal equivalent is estimated at 50 to 100 kg per person in the Sahel, compared with 7 500 to 8 000 kg in industrialised countries. In spite of this extremely low consumption level, Sahelian countries are now facing a major energy crisis. Apart from a few regions, there is a considerable shortage of easily accessible fuel wood. In many cases, the shortage is routinely met through permanent overexploitation of existing, accessible woodlands and the use of straw and plant detritus which can therefore no longer be used to fertilise the soil.

46

Forest products are used not only for domestic cooking and heating, but also for such activities as village blacksmithing. Fuel wood is also needed to process certain crops and to smokein fish. It is also used in industrial baking, tanning and ceramics.

Very little use is made of such positive factors as high-yield seed, irrigation water, and fertilizers, pesticides and herbicides. Only one farm in a thousand is mechanised.

Organic farming without chemical fertilizers or biocides, which uses composting and balanced crop rotations, improves soil fertility and could be used to advantage but priority biological control is non-existent or still at the stage of isolated trial operations. Biological farming using chemicals (fertilizers and biocides) is only practised in export cropping areas, covering on average 8 to 10 per cent of the territory.

Pesticide and herbicide use has increased slightly in recent years but the number of products available has expanded considerably, making any rational selection of a product even more difficult, especially in the absence of even partial regulations. This raises the issue of approval for pest control products. But toxicological studies require specialised laboratories, which in the Sahel do not exist or are still at the planning stage. The different types of test (behaviour, selectivity, residual effects) are not conducted satisfactorily.

Deteriorating Economies

All this means that the Sahelian economies are deteriorating, plagued as they are by unemployment, stagnation and insolvency. Deteriorating terms of trade are due at once to lower commodity prices (which have helped finance the development of the more affluent countries) and higher import prices (equipment, fertilizers, pesticides). Add to this the relatively high indebtedness level, which is rising everywhere as far as requirements are concerned, but chiefly as far as debt reimbursement is concerned.

In other words, Sahelian countries are faced with an impossible situation: the commodities on which their economies are based can no longer contribute to their development. In order to survive they have to borrow, with debt leading to further debts. In order to reimburse, they are "pushing" production, which leads to surpluses and dropping prices. Though this may benefit their creditors, it jeopardises their own development.

The outlook is not brighter for industry. Following independence, most governments aimed at public sector management of most industrial and tertiary activities. Today, most of these enterprises are facing bankruptcy due to overmanning, lack of motivation and disastrous management.

As far as agricultural policy is concerned, most countries, as we saw earlier, have seen production (with the exception of cotton) stagnate or drop. The factors promoting agriculture were not encouraged or insufficiently taken into account: producer prices did not provide incentives, under-equipment was and still is obvious, marketing networks and systems seldom fulfilled their functions, storage still remains to be organised, and the excessively cumbersome management of agricultural projects has considerably handicapped development. Last, but not least, the Sahelian countries have for too long relied on only one or two export commodities.

Remedial Action Taken

After assessing their respective situations, the countries concerned launched three types of action within regional co-operation bodies such as the Permanent Interstate Committee for Drought Control, known by its French acronym CILSS, and with the help of bilateral and multilateral co-operation partners.

First, studies were undertaken that aimed at improving knowledge of ecosystems and production systems, degradation processes, and various potential resources (water power, energy, etc.). These studies, which included surveys, sectoral analyses, meetings and consultations, covered such areas as irrigation and rain crops; ecology and forests; village and pastoral water supplies; livestock production; human resources, health and training; and recurrent expenditure.

With more comprehensive information about the Sahel available, it was possible to launch many programmes and help governments and co-operation partners determine their policy. Public development aid flows increased sharply, rising by 20.1 per cent per year on average between 1977 and 1981. Strategies are being drawn up for drought control and Sahel development. Guidelines are being prepared to set up action plans aimed at meeting the food and energy requirements of the population while maintaining the balance of the ecosystem and ensuring development.

Several regional co-operation systems have been launched. In 1973 the Sahelian countries of Western Africa set up a sub-regional co-operation body, the CILSS, with the objectives of promoting ways to control drought. It serves to centralise aid requests to donors.

In order to improve co-ordination with the United Nations system and at the Sahelian countries' request, FAO set up an Office for Special Relief Operations for the Sahel (OSRO) at its headquarters in Rome. OSRO was responsible for monitoring food aid consignments, co-ordinating cereal distribution in the region and collecting and processing data on emergency aid operations. It is no longer active.

With a view to co-ordinating assistance under the United Nations system, including technical assistance for project preparation and improving the coherence and co-ordination of aid from the international community, UNDP set up a provisional United Nations Sudano-Sahelian Office (UNSO), which is now permanent. It has been given special mandates, which have at times overlapped with CILSS activities, causing problems.

Following contacts and actions taken by volunteers, the OECD Development Assistance Committee set up the Club des Amis du Sahel in order to stimulate interest in long-term Sahelian development by public and private, national and international organisations; to organise the exchange of information between group members on the activities undertaken by them at the request of the Sahel governments; and to help prepare general and sectoral studies needed for the establishment of an outline or regional Sahelian development.

The Club du Sahel was provided with a small Secretariat (four officials seconded by France, the United States, Canada and the Netherlands) based at the OECD in Paris. Since 1976, the Secretariat has organised a conference every two years grouping Sahelian countries and bilateral and multilateral partners, leading to spectacular achievements in the understanding of Sahel problems, and in bringing together donors, the Sahelian countries and in increasing the inflow of aid. In spite of certain difficulties and criticisms, the co-operation and collaboration between the Secretariats of the Club du Sahel and the CILSS remain a unique and dynamic model of North-South co-operation which has helped, among other things, the Sahelian countries to remain optimistic and to be better understood by their co-operation partners.

The ACP/EEC Convention should also be mentioned since it has led to the establishment of financial and technical co-operation, the development of trade and services, and the creation of a mechanism for stabilizing export earnings (STABEX). It is a flexible and dynamic formula for North-South co-operation, with the advantage of adjusting to the different programming cycles.

Bilateral co-operation agencies (chiefly the United States, France, Germany, Switzerland, the Netherlands and Canada) have consistently supported the Sahelian countries and have succeeded in improving their intervention mechanism and aid level over time. Nevertheless, there is still room for better co-ordination.

Aid "transiting" through the United Nations system has encountered difficulties. Over 60 per cent of resources has been spent on studies and action has not always been as productive as one could have hoped. The impact of the different projects and programmes seems to be limited by such factors as the very scale of the drought and desertification, and the limited means to combat them. Add to this a lack of trained personnel and the absence of a suitable research policy, combined with a low level of involvement of the populations concerned. Activities have been highly sectoral and compartmentalised, and legislation is inadequate or non-existent. Lastly, co-ordination between co-operation partners has at times been poor.

New Proposals and Activities to be Reinforced

In order to reverse these trends, international technical co-operation must be improved and strengthened, while regional and national programmes must be based on or must integrate a number of closely linked action programmes.

This would consist mainly of various actions designed to create physical obstacles to the advance of the desert, to protect production areas, to restore spoiled areas and to prevent the gradual encroachment of sand on production areas, infrastructures and water sources, which calls for intensified dune-settling operations.

Soil conservation, long neglected in the Sudanese/Sahelian countries because of the extensive production systems and the availability of land, will be of greater concern in the years ahead, because of rising population pressure and the placing of new and increasingly vulnerable lands under cultivation. But soil conservation represents a substantial investment with significant effects in the medium and long term. It must be supported by the farmers and ought to be managed by them. Experience shows that investment in soil conservation is acceptable only when it is part of a more remunerative production system.

To bring quick results, priority must be given to developing and demonstrating soil conservation techniques in conjunction with the various types of regional development projects. They must also be developed for cotton and groundnut companies whose advisory and production facilities are present throughout wide areas.

Since ecosystems are fragile, any pastoral/agricultural development action must respect ecological balances and the production capabilities of the environment. It should aim at safeguarding and increasing natural resources: protecting existing resources by taking defensive measures or tackling the human problems, gradually winning back the broadest possible areas undergoing desertification in order to restore productive potential; conserving, improving and maintaining water resources both to raise the productivity of the soil and to provide populations with production resources that are guaranteed against climatic hazards.

Other goals are improving and developing the exploitation of resources, i.e. by developing the land to reflect its natural and human potential; developing village sites; and transforming production systems, especially by incorporating conservation and enhancement technologies for soil, water and vegetation, through systematic use of water resources and through introducing highly productive and drought resistant plant and animal species.

Equally important areas are controlling and organising the man/-environment relationship, i.e. by controlling demographic and urban growth; controlling and organising internal migration; and training and organising populations for the management and rational exploitation of their environment.

Co-operation partners can usually provide support with the following aspects: family planning, land use development, development of natural plant formations and research.

Development programmes should involve simple, basic initiatives that the local people can learn to manage, with decentralised planning focussed on:

-- Integrated and integral development (aforestation and agricultural activities properly backed up in terms of production factors, welfare actions, etc.);

-- Continuous monitoring of natural resources and their development;

-- The principle that any policy for counteracting desertification and deteriorating ecosystems must be harmonized with the attempt to achieve food self-sufficiency.

Partners could provide support in the form of inputs and equipment, satellite and aircraft image monitoring and help with local initiatives (protecting small areas, developing watersheds).

In the interests of a more efficient drive to rehabilitate and restore the environment, there is a need for legislative, institutional and communications measures. In particular, environmental protection and management legislation needs to be updated with better coding of how land is to be used in order to limit conflicts and overlapping in agricultural, forest, pastoral and fishery development projects.

The present rapid evolution of traditional rights to natural resources is marked by growing confusion over rights to the use of woodland/pastoral areas that are still collective. Several attempts have been made to renovate or to adapt new rural pastoral and forest codes, but the resources deployed for the purpose are in many cases totally inadequate and confined to a single nationwide code which may not be very suitable for certain regions. But any rural environmental policy can be applied only if there are suitable regional regulations, clearly defining the rights and duties of beneficiaries.

An appreciable investment will be required to have teams of lawyers, sociologists and technologists devise suitable new legal instruments. They will have to draft regional codes better adapted to the ecological and sociological background of pastoral and forest systems for exploiting natural resources as well as devise straightforward rules for resource management at the village level (e.g. rules for new village woodlands).

Partners in development co-operation could support communications and information actions: providing slide presentations, documentary films, producing and circulating magazines, scientific articles and informative brochures. Local language radio programmes can often be an effective medium, and have proved their worth in bush fire fighting operations. Television (which is making greater inroads with electrification) can play a similar part. Educational films and videocassettes are needed to explain natural resource management. Those can also be reinforced by travelling cinema shows which are always popular in villages.

Since the people of the Sahel are both the victims of and partly responsible for their deteriorating environment, producers must be organised and trained to make rational use of environmental potential and to restore the ecosystem. In other words they need to become actively involved in their own development, and not just onlookers. For this purpose a number of activities will have to be stepped up. Among other things, people will have to be organised into socio-economic interest groups, because only the active and collective commitment of those concerned can achieve tangible results; organisers and planners will need appropriate training and deployment, and there must be R&D promotion in environmental matters.

An energy programme

This requires precise objectives:

-- Rationalising the exploitation of wood by organising foresters and training them in forest management and charcoal-burning techniques.

The foresters will work to contractual specifications and in predefined areas; co-operation partners could provide technical support in identifying and inventorying areas for exploitation;

-- Reducing the consumption of firewood by widescale distribution of improved stoves; here the rural craftsman will need support both upstream, by providing him with the necessary inputs and assistance (metal sheets, among others) and downstream (marketing);

-- Stepping up research on alternative energy sources (biogas, solar, wind, etc.) with demonstrations of proven techniques for using them. Co-operation partners could bring their own know-how and advanced technologies to bear here.

Water supply programme

This programme comprises a set of activities which can be broken down as follows:

-- Updating inventories of groundwater and surface water resources and evaluating the uses to which they are being put, so as to more closely identify the further steps necessary to achieve a satisfactory water network;

-- Developing small farm plots in order to stabilize farming businesses and rehabilitate existing irrigation;

-- Monitoring and evaluating groundwater and activities to replenish it;

-- Building large numbers of small dams in order to limit water losses, replenish groundwater and safeguard crops.

Co-operation partners could assist in updating water resource inventories, developing and rehabilitating irrigation and in helping to exploit groundwaters.

Urban environment programme

Among urban environmental considerations, the planting of trees in towns and outlying areas is especially important since the climate is so harsh. Substantial efforts have already been made, often to create green belts around capitals or to plant tree-lined avenues.

Large scale investments remain to be committed in both capitals and secondary towns, to maintain recently established plantings; to co-ordinate new plantings with existing roadways and waste water drainage for shanty towns; to protect and renovate tree resources, often associated with historic town centres that are threatened by urban change.

These investments are often beyond the means of the local authorities.

Possibilities for strengthening co-operation are already identifiable, especially in the twinning of Sahel towns with towns in OECD countries. They could take the form of support for municipal maintenance services; participation in the financing of new projects; and initiating comprehensive schemes for tree-planting in towns, as part of town-planning studies in hand.

Agro-chemical input control programme

The use of pesticides, fungicides, herbicides and fertilizers has been growing, but the pattern has been uneven, even though these are areas where it is quite significant. More agro-chemical inputs is an essential step towards higher yields and food self-sufficiency, as can be seen in tropical Asia. They are, however, all designed, produced and marketed by only a few multinationals, that have been known to market products that are potentially hazardous to the population or to the fauna of Sahel countries. These multinationals could develop plant treatment and fertilizers that would be suitable for Sahel countries but they might not be considered sufficiently profitable.

The impact of agro-chemicals on the various component parts of ecosystems requires "downstream" research, which still remains to be undertaken in most of the areas in which they will be used. In practice, moreover, agro-chemical inputs will require a substantial advisory effort to ensure that recommended concentrations and safety precautions are observed.

All this suggests three lines of action:

-- Funding for manufacturers of agro-chemical inputs to encourage the development of products tailored to the environmental context of the Sahelian countries;

-- Training of national technicians, stepping up extension work in areas where agro-chemicals are already being used;

-- Funding for impact assessments in areas where these products are already being used and simulation impact assessments for areas in which they might be used.

Co-operation Programme

As Yves Berthelot of UNCTAD pointed out, multilateralism is going through a serious crisis. The western industrialised countries are tired of being put in the dock, and have become rather unreceptive to demands which the Third World does not always press so vigorously elsewhere. The developing countries are disappointed to see that so many negotiations are making no progress, or lead only to oft repeated resolutions that are never implemented.

As regards the multilateral institutions, the industrialised countries naturally have a preference for such institutions as GATT, the IMF and the World Bank. However, they are probably right to suggest that the multilateral system allocates excessive resources to expert reports, seminars and their administration at the expense of concrete activities.

Bilateral co-operation, on the other hand, presents disadvantages for the developing countries: long lead times for project review and disbursement of resources, so that by the time a project is actually implemented the original conditions often have changed. (This is what often accounts for setbacks). Bilateral co-operation also tends to take a very sectoral approach, with little or no integration of individual activities, while at the same time the consultancy and study phases tend to be unduly burdensome.

Non-governmental organisations have drawbacks, since most of them suffer from such weaknesses as lack of appropriate technical skills and know-how, as well as surprisingly heavy operating costs.

The Way Ahead

Environmental development can be tackled in two ways:

-- In the short and medium term, the existing structures of the rural world should, as suggested in the ACP-European Community Courrier of June 1985, be regarded as a stock on which to graft a succession of activities to promote, accelerate, entrench and extend the development process. Taking the structural characteristics of the rural world into account means trying to look at it comprehensively, because a strongly integrated peasant society must be seen as a whole.

-- In the longer term we need to refine our understanding of the rural world in terms of its future needs, so as to devise strategies with which to meet them. This means identifying the external factors which are breaking down core groups and structures, along with the forces underlying the emergence of new roles, new kinds of status, new decision centres, new forms of social and geographical mobilisation. In acting on all this, our co-operation partners should also have regard to the overwhelming need to cut down on red tape, as well as on the costs of consultancy and studies.

To paraphrase the Wall Street Journal, the best thing we can do for indebted countries and for developing countries in general is to clarify our own ideas of what is necessary for development to succeed.

B. THE SEARCH FOR WORKABLE SOLUTIONS

by
Glenn Slocum
Club du Sahel, OECD

The Club du Sahel was set up in 1976 to support the work of the Permanent Interstate Committee for Drought Control in the Sahel or CILSS to promote food self-reliance and to reverse environmental degradation in the drought-stricken countries of West Africa (the member countries of the CILSS are: Burkina Faso, the Cape Verde Islands, Chad, Gambia, Guinea-Bissau, Mali, Mauritania, Niger and Senegal). These countries have fragile ecosystems and are especially vulnerable to climatic change. They are generally character-ised by low annual rainfall patterns, with zones ranging from desert and Sahel-Saharan in the far north to Sahel-Sudanian and Sudanian zones in the south.

These countries face a major challenge: how to increase the produc-tivity of their natural resources without degrading them. Environmental degradation is caused by a mixture of natural and human factors: lack of rainfall, increased population pressure on the land, overgrazing. Studies have shown, for example, that grazing regions in the northern zones of the Sahel cannot support a population density exceeding 0.3 km2, but the current average is 2.0.

In the Sahel-Sudanian zones further south, where a combination of herding and agricultural activities takes place, a World Bank study indi-cates that the maximum carrying capacity of the land cannot exceed an average of 15 people per km2, whereas the average now is more than 20, with much higher densities in certain zones such as the groundnut basin of Senegal, the Mossi plain in Burkina and the Zinder-Maradi region of Niger. The result is devastation of the natural brush cover, loss of forests, insufficient fallow and soil erosion.

In collaboration with the CILSS, in the period 1979 to 1983 the Club undertook a number of sector appraisals country-by-country, notably in irrigated and rainfed agriculture, village water supply, ecology/forestry, fisheries, and human resources. The livestock sector is currently being analysed in certain of the countries. These appraisals generally deal with the array of constraints to production, including inappropriate policies, socio-economic factors, limitations on natural resources and institutional problems.

The ecology/forestry reports showed that fuel-wood demand represented the major pressure on forest resources and was, therefore, a leading cause of land degradation, while government policies often undermined plans to increase forest production. Donors accorded (and continue to accord) scant priority to this sector. A number of projects introduced new species which were expected to be faster-growing and more productive, but their performance was often about the same as that of indigenous species.

Costs of forestry projects were high, and, therefore, unsustainable by national forestry services or the local populations. In fact, most projects were being designed from the top down, and complicated land tenure issues combined to discourage villagers from expanding their tree-growing. The studies pointed the way, however, to some necessary reforms, among them in land tenure, popular participation, natural forest management and agro-forestry. We are subsequently finding that, while government-to-government projects (the traditional donor approach) are encountering implementation difficulties ranging from moderate to severe, there is mounting evidence of some real successes at the local level, usually encouraged by the intervention and support of outside or indigenous non-Government organisations (NGOs).

Everyone agrees that the trend towards land degradation (sometimes referred to as desertification, a term that does not meet with unanimous consent in the scientific community) must be reversed. The challenge is to find and adopt the appropriate mix of measures which are implementable. This report highlights our lessons and findings over the last 11 years and suggests some guidelines for the future.

What Are the Constraints?

Population growth

It is estimated that, at current growth rates, the population of the Sahel will be 40 million by 1990 and 50 million by 2010. With clear evidence that land degradation is on the rise due to the fact that more farmers are cultivating marginal land as well as to poor rainfall patterns in most years and overgrazing by animals, Sahel countries must develop coherent national population policies and land-use plans.

Methods of conceiving development programmes and assuring popular participation

Everyone agrees that sustainable development activities must be fostered. But the practice remains the same: governments agree on development priorities, collaborate on investment plans, obtain funding, and try to impose schemes which too often cannot be sustained by the target population. Development programmes must revolve around these guiding principles: sustainability, replicability, incentives (i.e. there has to be a pay-off for the producers, which means economic, financial and social soundness), and technical appropriateness.

Finding the motors of development

Due to the fragile ecosystems of the Sahel countries, their economies are inextricably linked with the more productive coastal countries. This is reflected in a number of different indicators, such as Sahelian immigration and cross-border trade. This argues for more regional planning and concertation. Unfortunately, the panoply of African regional organisations has not provided the hoped-for coherence to this planning process. It is essential that future development plans be based on the real productive potential of the countries, not on unachievable goals.

For instance, agriculture should be encouraged in areas where the water resources and soil quality justify investments. In the absence of adequate water supply, alternative investments, such as livestock, should be considered.

Inadequate planning/management capacity and insufficient operating resources

Government bureaucracies are confronted by a host of conflicting problems. Revenues are down as a result of declining investments and incomes (because of recent drought conditions, among a host of reasons). Government services are under-funded and numbers of skilled managers and technicians are limited. This means fewer services are available to the population, both in terms of advisers and goods, such as subsidies and credit.

While training programmes can provide more skilled service personnel, budget constraints mean that there will be limited possibilities for improvement in the foreseeable future. This in turn means that programmes must be designed for management and implementation largely by the target population themselves, such as village groups and co-operatives. These institutional constraints are as serious as the other constraints in dealing with environmental problems because they can block the entire process of defining, planning, managing and implementing a project.

Land tenure

Existing land tenure laws have been widely studied and the conclusions are similar: if ownership is not clearly defined, land users do not have the incentive to produce. This is especially problematic with respect to tree-growing. Agricultural lands are usually controlled by some kind of village organisation, whereas legislation over trees does not typically guarantee harvest to individuals. The introduction of agro-forestry has helped but improvements in the existing land-tenure arrangements are urgently needed. Governments' failure to act on this is, at least in part, due to the institutional weaknesses identified above.

Insufficient donor co-ordination

Here, too, everyone advocates closer donor co-ordination, sometimes called concertation. But diverse planning and programming procedures militate against full compliance with this goal. The Club du Sahel has learned that different mechanisms can fulfill different purposes. In-country co-ordination can take a variety of forms but sector-level issues can be addressed regionally as well. For example, the World Bank Consultative Groups and UNDP round tables help focus attention on the macro-economic issues and development planning goals.

The Club has created a number of informal, sector-specific groups made up of Sahelian and donor experts who, in sharing their knowledge and experience, reflect on progress, problems and achievements. Despite these measures, effective co-ordination of a country's investment programme remains an elusive goal.

Encouraging realistic objectives

This problem is really a function of the planning, management and donor constraints described earlier. Donors and host governments are too often seduced into financing schemes which do not meet the criteria of success. Shelves are stacked with reports of projects which failed to meet their objectives because they are not socially, economically or technically feasible. Some especially vulnerable areas to this kind of error have been large-scale irrigation, industrial forest plantations, and green belts, but there are many others.

The Club is playing a major role in getting donors and Sahelians together to debate the issues and, together, to improve their programmes. Many observers believe that, contrary to the frequent appeals for increased development assistance levels to Africa, the real problem is an insufficient number of carefully-prepared, appropriately-conceived projects within the context of a coherent development plan which meets the concerted approval of donors.

What Have We Done?

Sector analyses

The ecology/forestry sector analyses, carried out between 1981 and 1984 for most of the Sahel countries, helped focus attention on the progressive de-forestation of the region and on the decline in soil quality. A synthesis of these reports (issued as Sahel D(83)194, November 1983) shows that in the 1970s only about 1 per cent of official development assistance (ODA) went into the forestry sector.

There was only slight improvement on the percentage of allocation in the early 1980s but after 1982 the figure slipped below 1 per cent. The paucity of resources devoted to the forestry sector reflects the difficulty of identifying successful projects. However, the lessons are clear, and the synthesis report offers a number of recommendations for action, among them:

-- Introduce a system of tenure that will benefit villages, districts, families or individuals;

-- Transfer as much authority as possible to the local level, for example, to village councils;

-- Integrate the tree into the agricultural and pastoral environment;

-- Rewrite existing forestry legislation to conform with the demands of a system of collective and private ownership;

-- Reduce consumption of firewood and charcoal;

-- Protect and develop natural woodlands.

Efforts to move in these directions have been initiated, but success has been spotty. Reform of forestry and rural codes and decentralisation of power are under consideration in some countries but have not yet been implemented. Most countries have enacted legislation to protect forests and reduce fuel-wood consumption but provision of incentives to encourage tree-planting is constrained by the problem of thinly-staffed government services and low budget resources.

Regional seminar to develop a desertification control strategy

The CILSS organised a seminar in November 1984 in Mauritania, which brought together for the first time all the Sahelian directors of agriculture, livestock and water and forests, as well as donor experts from all the major bilateral and multilateral agencies and some NGOs. Together they developed a general strategy, which is essentially a set of principles to follow.

These include the need to re-orient activities to render the target population responsible for them. It calls for such strategies to be integrated into overall national development plans and, most importantly, it calls for multi-sectoral approaches to preserve the natural resource base and rehabilitate its productive potential.

Preparation of national desertification control plans

With assistance from CILSS and the Club, several governments have put together what they call national master plans for desertification control. They typically contain three sections: the first summarizes the current state of the environment in the country and describes the causes of degradation. The second outlines a proposed strategy and programme orientation. The last section, which the countries usually add, is a "shopping list" of projects for funding by the donors. (This list is often not correlated with the previous sections.)

National meeting with donors to review the plan

Four countries have presented their plans to donors: Niger, Mauritania, Mali and Burkina Faso. In two cases, inter-ministerial co-ordination committees on desertification have been established, which meet periodically with donors.

However, evidence to date suggests that these mechanisms have not really provided the impetus for effective changes in approaches to natural resources management.

This is probably due to a number of factors:

-- The plans often reflect the work and view of only one department or ministry, typically called the ministry of environment, or of water and forests, or protection of nature. In many cases, the meeting with donors revealed a lack of sufficient input and accord from fellow technical services, such as agriculture and livestock, or from central ministries such as planning and finance;

-- Follow-up and monitoring of the decisions made at the meeting are insufficient, as evidenced by lack of co-ordination between the technical ministries, despite the establishment of inter-ministerial committees;

-- The plans lack the depth of analytical substance required to set the stage for project re-orientation and/or identification. This indicates the need for further work;

-- Donors cannot re-orient their programmes overnight. They need to gather more information on the nature of the problems and spend greater effort, in collaboration with the host countries, on identifying workable solutions.

What Do We Propose to Do?

Identify successful projects

-- A number of non-governmental organisations are working on relatively small-scale activities in the field, which are showing positive achievements, i.e. developing sustainable production systems which safeguard the natural resource base. Some major donor projects are also showing promise. There have been some recent attempts to highlight these successes.

-- The International Institute for Environment and Development (IIED) sponsored a Conference on Sustainable Development in April 1987 in which four successful projects were analysed and presented.

-- The World Bank is conducting an internal assessment of all nature protection activities in certain countries of West Africa.

-- The World Resources Institute is proposing to undertake development of a co-operative programme to promote resource conservation and agricultural productivity in sub-Saharan Africa.

-- The German CILSS Programme (Programme allemand CILSS, headquartered in Ouagadougou, Burkina Faso) is developing innovative integrated resource production projects in selected ecological zones in the Sahel.

Conclusions

The relationship of the Sahelian population with its environment is no longer in balance. Traditional grazing land has been lost, agricultural land has become less fertile and the forest cover has substantially diminished.

Drought has lowered rainfall averages while the number of Sahelians has doubled in a generation, bringing with them a significant expansion in the application of production systems. The restoration of previous, more favourable rainfall patterns may bring a temporary balance but will not be sufficient to offset these influences.

Donor countries must work more closely with developing countries to identify positive measures and undertake to expand them. To succeed at this, the following conditions must be met:

-- A willingness of host countries to organize themselves more effectively and to abandon costly, grandiose projects in favour of smaller projects which provide decentralized authority to local population groups;

-- A willingness of donor countries and organisations to re-orient their programmes accordingly;

-- The commitment of both to experiment and capitalize on limited successes.

FOOD SECURITY AND PEST CONTROL:

A. THE GOAL OF INTEGRATED PEST MANAGEMENT

by
Edward Johnson
US Environmental Protection Agency

The race between food production and population growth has resulted in a tremendous intensification of agriculture and the adoption of modern technology to increase yields and improve crop quality. However, the components of modern agriculture, such as continuous cropping, monoculture, increased fertility, irrigation, and new high-yield crop varieties often increase the vulnerability of crops to pest attack. FAO estimates that about one-third of the world's food crops are destroyed by pests during production, harvest and storage. In developing countries, it is estimated that the figure may be as high as 50 to 60 per cent annual losses. Unless measures are taken to protect food crops against the ravages of pests (insects, weeds, plant pathogens, nematodes, rodents, etc.), the production gains of the new technology will be lost.

The need for a systematic management approach to crop production, including crop protection, based on sound economic, technical, sociological and environmental considerations is critically needed. An integrated pest management approach offers the greatest promise for an effective, safe and continuing solution to many pest problems. Integrated approaches have been implemented successfully with the best documented cases in developed countries. While a number of pest control measures are available, developing countries rely almost exclusively on pesticides for their pest control needs.

Pesticides are essential tools in pest management but they also have the potential to produce adverse health and environmental effects. The impacts of pesticide use in developing countries have major implications that extend beyond their own borders. As more agricultural commodities are exported from developing countries to industrialised nations, importing countries have become more concerned that pesticides be properly used by exporting countries to prevent accumulation of excessive levels of residues on commodities entering international trade. In addition, the indiscriminate use of pesticides has also resulted in accelerating pesticide resistance, a phenomenon that threatens food production and public health worldwide.

Most of the active ingredients used in pesticides and formulated products are imported by developing countries. While developed countries have the necessary infrastructure to regulate use of these products, many developing countries do not. Trade in pesticides from industrialised countries to developing countries continues to increase. After 40 years of extensive promotion, pesticide markets in the developed countries are saturated; but the demand is still rapidly growing in developing countries. The environmental and health hazards posed by the use, storage and disposal of pesticides in these countries has increased, and to address the issue, many governments in developing countries have initiated regulatory programmes to control the risks involved.

There is a continuing need to balance the benefits and the adverse environmental impacts of pest management practices in the development co-operation context. Four objectives should be borne in mind:

1) To improve communication and co-operation between donor agencies and developing countries with respect to pest management;

2) To consider policies and approaches to improved pest management;

3) To explore approaches to environmentally sound pest management in developing countries;

4) To identify environmental consequences of inappropriate pest management techniques.

Pest Management Tactics

Pest management is a multidisciplinary approach to plant and animal protection based on economic, social, ecological and technical factors. The protection of crops and human health against pests is critical as world population expands and the demand for food increases. It is estimated that one-third of the world's food production is destroyed by pests. The amount is even greater in developing countries. Ironically, the same components that contribute to increased production under modern agricultural technology also increase the incidence of pest problems on crops. There is a need to control these pests to preserve the production gains of modern agriculture.

The direct control tactics available to combat pests can be summarised as follows:

-- Cultural control practices consist of manipulation of pest populations by selected traditional methods employed in crop cultivation;

-- Biological control involves the regulation of pest populations by their natural enemies;

-- Selection and breeding pest resistant varieties is one of the most economically important and environmentally sound pest management tools available;

-- Use of pesticides provides a rapid, convenient and dependable means of controlling whole complexes of major pests. Properly used, pesticides are essential elements in a pest management programme.

Dependence on a single pest management tool has often proven to be short-lived and, in many instances, to have unwanted side effects. Pesticides will continue to play an important role in pest management, but an integrated approach combining the use of all available technologies offers better prospects for long-term management and reduced environmental damage.

Integrated pest management (IPM) is defined as a "pest management system that, in the context of the associated environment and the population dynamics of the pest species, utilises all suitable techniques and methods in as compatible a manner as possible and maintains pest populations at levels below those causing serious economic injury." It captures the advantages of some control practices while avoiding the disadvantages of others. It keeps the cost of pest control to a minimum and improves pesticide efficiency and is thus economically and environmentally sound. IPM reduces adverse environmental and public health impacts by considering the economic, social, and environmental effects of available pest management options, and it also increases profitability.

Constraints to integrated pest management implementation

Successful field level implementation of IPM has been well documented in the United States and Europe, but there are fewer examples of successful IPM in developing countries. The IPM approach requires considerable research, highly effective information delivery systems, and an appropriately trained extension service, facilities which many developing countries often lack.

Of the various components of modern agriculture, IPM presents by far the most difficult challenge to traditional, small-scale farmers in developing countries. IPM requires the farmer to grasp a far more complex set of data to form the basis for his day-to-day decisions on pest management. IPM farmers must achieve greater discipline in improving their pest management practices, including more frequent and precise attention to pest identification, danger symptoms, timing and anticipated returns to costs incurred. Concurrently, we must learn from the farmers' wealth of experience incorporating their knowledge into IPM programmes.

The major constraints in developing countries to successful IPM implementation are as follows:

-- IPM involves complex technological principles requiring intensive training and education of farmers. In order to implement IPM effectively, one must: have an understanding of the biological and environmental factors associated with plant and animal growth; know what plants need and provide proper cultural care; and be familiar enough with pests and their life cycles to apply the best controls correctly and at the right times. Farmers must learn the principal natural enemies and methods of estimating their relative strength. They must gain command over economic threshold level concepts, selection of the correct control measures, and timing and application rates. The approach should be simplified to make it more acceptable and easier to adopt by small farmers;

-- The development of resistant varieties has not kept pace with the rapidly changing pest complex situations. In Asia, for example, resistant seed varieties have been developed mainly for rice and almost exclusively for irrigated systems. Resistant varieties for rainfed production and for other important crops have yet to be developed;

-- The economic threshold concept, one of the basic components of IPM, is difficult to apply in the field. The IPM approach is based upon the concept that there is a pest population level below which the cost of applying control measures exceeds the losses caused by the pest. This level is called the "economic threshold." When this threshold is exceeded, it becomes economical to initiate control measures. IPM does not attempt to eradicate all pests, but rather to keep damage below these levels. It is essential that the economic threshold be accurately established by proper measurement of pest population density. Inaccuracies will lead to unnecessary control measures and/or excessive crop losses. Extensive research has to be done to estimate the economic threshold levels for each crop-pest situation, and in many developing countries, IPM technology cannot progress because these levels have yet to be defined. In other situations, research has provided the necessary information but the transfer to straightforward field use has not been completed. This gap can best be filled by well-trained extension personnel;

-- Surveillance procedures for pest damage are complex and laborious. Regular monitoring or scouting procedures are designed to detect natural enemies, identify pests, and determine the levels at which pest populations reach the economic threshold levels. A thorough assessment of the problem and the risks involved leads to efficient pest management decisions. Doubts have been expressed on the accuracy of data gathered under current monitoring systems. Surveillance by farmers of their own fields seems to be a more effective approach to consider, but this requires extensive education and training;

-- The limited financial, institutional and technical resources available in developing countries has hampered the adoption of IPM on a large scale basis. The overall success of IPM depends to a large extent on the technical support in terms of research, education, and extension programmes of governments. More resource efficient alternatives to traditional research, education, and extension programmes need to be developed and tested for each programme to be achieved in many developing countries;

-- Country policies both in LDCs and in exporting countries which subsidise pesticides encourage pesticide use at the expense of more balanced IPM programmes.

Pesticide Use

Although IPM is the preferred pest management approach, because of the current difficulties described above, many developing countries rely exclusively on pesticides as a convenient and reliable control strategy. In many cases, plant protection policies are confined to the choice of pesticide products, their dose rates, and proper application techniques.

Many of these countries have enjoyed the benefits of pesticide use in terms of increased yields and protection of health. While demand for pesticides is levelling off in the industrialized countries, the use of pesticides is expected to increase in the developing world. In Asia alone, the pesticide market has shown an average annual growth of 10 per cent during the past five years and the trend is expected to continue in the foreseeable future. On the whole, insecticides form the bulk of the pesticides used in developing countries, and many of these are the less expensive and more hazardous products. Although DDT, BHC, and other organocholorines are banned for agricultural use in almost all the developed countries, they are still the most extensively used pesticides for public health programmes in these countries. For example, in the Asia/Pacific region, DDT and BHC rank among the top five pesticides on the market.

Although the trend in the pesticide industry points towards development of pesticides which are effective at lower doses, more target-specific and less hazardous, developing countries have been slow in adapting to these changes due mainly to economic factors. The cost of pesticides continues to increase and, in most cases, as pest outbreaks become more serious, the dependence on more traditional, less expensive pesticides may be the only immediate solution for small farmers to save their crops.

Significant environmental and health concerns

The increased use of pesticides has raised a number of concerns. Among the more significant environmental health programmes associated with pesticides are the following:

-- Health hazards. The potential of pesticides to produce adverse health effects are well known. Most chemicals used for pest control are not only toxic to the intended target organisms, but also to man. Hazards due to indiscriminate and prolonged use of pesticides have been a primary concern of all countries. Studies show that the rate of pesticide poisonings in developing countries is at least 13 times as high as that in the industrialised countries. Although developing countries use only one-fifth of the world's pesticides, they suffer half the poisoning cases and nearly three-quarters of the deaths;

-- Residues in food. Pesticides applied to protect against pests in the field and during storage may remain on a crop for some time after application. Consequently, concern is often expressed about the level of residues that remains on foods that are sold and consumed. Developing countries are now concerned about the residues on export crops which can result in their refusal by the importing country. Although governments have taken some steps to control the levels of residues and protect consumers, many developing countries lack resources and sufficiently trained scientists to devote more attention to residue regulation;

-- Environmental effects. Only in recent years has the extent of environmental contamination from pesticides become clear. Reports of the deterioration of fish and marine life, adverse effects on wildlife and livestock, and contamination of bodies of water traced to pesticide use have caused concern to many governments. But because of the lack of resources, many developing countries have no means of carrying out routine monitoring of such adverse effects.

-- Pesticide resistance. Pesticide usage has the potential to induce resistance in pest populations and thus undermine its effectiveness as a means of pest control. Pesticides destroy natural enemies that keep pests under control. In addition, some pests have a natural tendency to become resistant to continued use of the same chemicals. The common response of farmers to resistance problems is the use of larger doses of pesticides, which can lead to more serious problems. Accelerating instances of pesticide resistance represent a major worldwide problem that is having an increasingly severe impact on pest control and agricultural production.

How Policies Affect Farmers' Choices

Farmers' choices of pest management techniques are determined by a number of factors: training, information and education; availability; pricing and credit policies; and regulations and controls. These can be influenced by government policies and programmes.

While many developing countries have started to implement some IPM approaches, plant protection policies are still generally confined to the choice of pesticides, their dose rates, and proper application techniques. The goal of such policies is to ensure maximum benefits from pesticide use while minimising the risks involved. Among the general principles to be adopted are:

-- Proper selection of pesticides. The selection of pesticides should be based on their toxicity to specific target pests. Pesticides that are known to harm important beneficial species should be avoided. The traditional practice of using broad spectrum products that control several pest species with a single application has resulted in the elimination of a number of beneficial insects. However, these negative effects were balanced against the benefits of cost-effectiveness. Modern agriculture takes all these into account and encourages the use of highly specific pesticides;

-- Timing of applications. Many of the pesticides available in the market can be used in ways that enhance their ecological selectivity. There are a number of ways to achieve useful levels of ecologically selective action. Applications should be timed to coincide with the most susceptible stage of pest development. The proper attention to timing of applications can assure effective control of pest species with minimum amounts of pesticides and the least possible adverse effects on natural enemy populations;

-- Placement of pesticides. The more precisely and strategically pesticides are placed on their target, the lower are both the level of environmental pollution and the cost of application. Examples include treatment of borders of fields to control pests migrating from outside areas; treating seeds instead of blanket treatment of entire fields; control of locusts by spraying widely spaced strips of vegetation in the path of hopper bands, etc.;

-- Proper application equipment and techniques. Not all pesticides applied actually reach the target site. The ability to treat only certain portions of a plant with a pesticide could greatly reduce the quantities needed for acceptable control. The behavior of many pests is that much of their activity is restricted to certain portions of the host plant. Proper equipment and application techniques are crucial for the proper use of pesticides. Proper application techniques can increase the effectiveness of the pesticide and minimise hazards;

-- Appropriate pesticide formulation. The manner in which a pesticide is formulated has a strong influence on its effectiveness and the safety of its application. The type of formulation affects shelf-life, compatibility with other chemicals, reduced toxicity to non-target organisms, persistence of residues, uniformity of particle size, and convenience of application;

-- Proper dose rates. It is possible to select a dose rate that will control pests within acceptable levels and allow the natural enemies to survive. Selection of proper dose rates can also reduce resistance problems and improve effectiveness of pesticides.

Training and Information Dissemination

To support these policies, training and extension programmes geared towards educating users on the safe application of pesticides have been undertaken in many countries. Along with these programmes, information and mass media campaigns on pesticide safety precautions have also been launched. However, until an evaluation measure is available, such as statistical poisoning surveys, the impact of these programmes can hardly be assessed.

Training programmes on identification of pest species and natural enemies, rearing of beneficial insect species, and surveillance of pest populations have been initiated in many countries but progress has not been as extensive as those involving pesticide safety.

Regulatory Policies and Programmes

To further strengthen the controls on the use of pesticides, many governments in developing countries have instituted legislation to regulate the importation, production, distribution, sale and use of pesticides. Pesticide regulations cover a number of aspects, such as:

-- Registration, i.e. the process of evaluation and acceptance by the appropriate government body of documented proof of claims for efficacy and safety made for a proposed pesticide. The purpose of registration is to ensure that pesticides, when used in accordance with the label directions, will be effective and efficient for the purposes claimed and will not subject the user, the consumer of treated foods, and the natural environment to unacceptable risks.

The process of registration is complex and requires considerable expertise in various scientific disciplines. Developing countries often lack the resources and trained personnel to carry out the rigorous work of pesticide registration. In most cases, developing countries rely on regulatory decisions in developed countries, notably the United States, Germany and Japan. There are limited capabilities to assess decisions on pesticide use on the basis of country needs and priorities.

FAO has developed a set of guidelines to assist countries in implementing the registration system. For countries with no existing legislation, the International Code of Conduct for Distribution and Use of Pesticides can serve as a useful basis. There is a need to inform governments of these guidelines and of the Code and monitor their implementation;

-- Regulation of labels and advertising. Almost all countries regulate label requirements regardless of whether a formal regulatory system exists or not. Requirements usually include the claimed product specifications, directions for use, and precautionary measures. Many countries have adopted a colour coding system based on hazards classification of pesticides. This facilitates farm-level information on the pesticide toxicity;

-- Regulation of domestic production. All developing countries with a regulatory system have provisions to control the manufacture and formulation of pesticides. Government policies differ depending on priorities. Some developing countries encourage local manufacture of active ingredients to lessen dependence on imports and preserve foreign exchange. In many cases, regulatory policies tend to protect these industries. In most instances, countries import the active ingredient and formulate the finished products locally. In all cases, regulations exist to protect workers in these factories but their effective enforcement remains a problem;

-- Quality control and regulation of residues. One of the most common problems in developing countries is the proliferation of pesticides from local formulation and repackaging outlets. Many governments have no facilities to monitor the quality of the products. While consumer protection laws exist, monitoring facilities are not available and enforcement is extremely difficult. Because of the high cost of pesticides, farmers are concerned about the quality of the products they buy.

Pesticide residues are a problem especially in export crops. Importing countries impose legal limits on residues which if exceeded make a shipment unacceptable. Exporting countries are becoming more aware of the implications of shipments containing excessive pesticide residues and have begun to require the monitoring of export crops. However, because of limited resources and trained analysts, residue monitoring of crops produced for local consumption has been given low priority.

-- Regulation of imports/exports of pesticides. All countries with a
  regulatory system have a scheme to control pesticide imports and
  exports. The policy is the same everywhere: no pesticide can be
  imported or exported unless it has been registered in the country.
  In many cases, an import permit from the appropriate agency is
  needed before a pesticide can be imported or exported. Policies on
  banning and restricting pesticides are most effectively implemented
  through these controls on imports. In most cases, decisions abroad
  influence these policies. It is therefore important that when an
  exporting country restricts a particular pesticide it notify the
  importing country of the action taken. This gives the government of
  the importing country an opportunity to assess the risks associated
  with the pesticide and to decide on the importation and use of the
  product concerned, taking into account local public health,
  economic, environmental and administrative conditions.

   Many other factors indirectly influence the pesticide regulatory
policies in developing countries, such as the political structure of the
government, economic needs, scientific and financial resources, etc. These
must be considered when reviewing each country's regulatory situation.

Pricing and Credit Policies

   Governments also influence the choice of pest control techniques used
by farmers through pricing policies for agricultural inputs and agricultural
commodities. In many countries, governments subsidise the use of pesticides.
These subsidies are provided directly in the form of below cost prices for
pesticides, exemptions from consumption or sales taxes and relatively low
import tariffs. Indirectly, sales are subsidised by access to favourable
foreign exchange rates and access to below market credit.

   The rationale for pesticide subsidies as well as subsidies for other
agricultural inputs is generally to reduce the risk to farmers of adopting new
technological packages such as those associated with new high yielding seed
varieties. In some cases, agricultural inputs are kept artificially low to
compensate for the low returns they earn when selling commodities locally, for
example, rice whose prices are controlled to keep food prices low for urban
consumers. In other instances, subsidies are granted for the use of locally
manufactured products in an attempt to stimulate local industry.

   Since farmers must balance the costs of crop losses against the costs
of alternative pest control techniques, any distortions in the markets for
agricultural inputs or outputs affect farmers choices of pest control
techniques. Frequently, subsidies lower the cost of pesticide use relative to
other control techniques and may encourage excessive or inappropriate
pesticide use. Subsidies also represent substantial lost revenue and direct
budget outlays for governments.

B. A STRATEGY FOR AFRICA'S SUBSISTENCE FARMERS

by
Dr. Eliud O. Omolo
Senior Research Scientist/PESTNET Co-ordinator
International Centre of Insect Physiology and Ecology (ICIPE)

Perhaps the gravest crisis Africa must face today is food security, but the problem is compounded by heavy grain losses caused by various insect pests. To date, the principal control methods have involved the use of chemical pesticides, but as experience has shown, these have proven both problematic and disappointing. For one, the majority of the resource poor farmers in the developing African countries have small holdings and lack adequate resources or the technical know-how to use pesticides. Moreover, the indiscriminate use of pesticides has brought in its wake insect resistance to chemicals, secondary outbreaks of pests, a rapid resurgence of sprayed pests, toxic residues, direct health hazards from pesticides as well as heavy economic costs

The availability of meat and other animal products for human consumption in the Third World has declined in recent years not just as a result of the slow growth in production and the rapid increase in human population, but also because of livestock diseases transmitted by ticks and tse-tse flies, including East Coast Fever (Theileriosis), anaplasmosis, babesiosis and trypanosomiasis.

To control these pests, farmers and governments are faced with a choice: not between pesticides and no pesticides, but between a hazardous and potentially ineffective or counterproductive pest management stragegy based on uncontrolled application of chemicals, or a pest or vector managment strategy which is safer and more effective. Such a strategy would incorporate farmers' traditional practices instead of replacing them and would apply technologies that could safely be used. This would reduce heavy dependence on imported inputs and would be focussed on the needs of rural farmers.

Pesticides, especially DDT, used in the control of mosquitoes and agricultural pests, have generated serious pollution of soil and water. Excessive use of herbicides in the Aswan dam reservoir to control weed infestation has caused adverse side effects on irrigated agricultural land and has also reduced fish production where Nile River water is discharged into the sea.

Right now in East Africa, the fish are dying in Lake Victoria, and scientists are working round the clock to establish the cause in order to develop control measures. The construction of a dam on the Zambezi river created an extensive area of marshy habitat in which the tse-tse fly population built up, increasing human and animal mortality in the vicinity. Therefore, a systems approach and interlocking resources management mechanisms are necessary.

ICIPE's Pest Management Network for Subsistence Agriculture in Africa

The International Centre of Insect Physiology and Ecology (ICIPE) was created to meet the need for alternative pest control strategies. Its mandate is research in integrated control methodologies of crop and livestock insect and arthropod pests, as well as the control of insect vectors of tropical diseases crucial to rural health, especially in Africa. One way in which ICIPE is planning to strengthen environmental co-operation among developing countries in Africa is through a Regional Pest Management R&D Network (PESTNET) for integrated control of crop and animal pests.

Work done at ICIPE, national research programmes and international research centres thus far has provided a knowledge base and has developed methodologies which could be utilised in the proposed pest management network. For example, ICIPE has generated information on plant resistance to several major crop borers, as well as components of intercropping and biological control which may be validated and introduced in the network. A number of potential pest management strategies based on the results of the work on these aspects by ICIPE hold promise for controlling the target pests. The dominant pest at any particular site and season might vary, so that the pest management system selected would need to be appropriately modified as a result of field experience, and combinations of components be integrated into a single package tested and modified with a view to assessing effectiveness, practicality, social acceptance and environmental concerns.

The Secretariat of the African Regional Pest Management Research and Development Network (PESTNET) will be based at the ICIPE Headquarters in Nairobi, and through an interactive network, the countries initially participating will establish related experimental pest control programmes. A documentation facility will be established at ICIPE to promote information exchange and dissemination. In addition, short courses in pest control methodologies will be conducted. Annual meetings of the participating countries will be held to undertake evaluation and periodic review of progress as well as to assure unity of purpose and regional commitment to the success of the pest control strategies adopted.

Although the project will be executed by ICIPE, the work plan will be undertaken utilising the existing knowledge base in national programmes in Africa, at ICIPE and at other international institutions. Apart from the links already established by the ICIPE with these programmes, PESTNET cognizance will be given to the relevant regional and country programmes already established by the Food and Agricultural Organisation of the United Nations (FAO) and other regional organisations.

We are quite confident that we have some knowledge that might lead to the management of tse-tse borne diseases for we have observed that the tse-tse fly can be attracted from a great distance and over periods of days or even weeks by an attractant naturally found in buffalo urine. We believe our research could lead to the evolution of a tse-tse fly "super-trap". Another of our scientists has observed that the fly carries a specific virus which causes sterility in both male and female tse-tse flies, particularly the latter.

We are also utilising a nematode (worm) which controls insects and which also carries a poison-producing bacterium that can independently control the insect. The Kenyan research team thus believes it may have found a "natural" way of tackling one of Africa's greatest scourges.

Funding for this project is badly needed. The Director of ICIPE, Professor Thomas R. Odhiambo, has attended meetings of the Technical Advisory Committee (TAC) of the Consultative Group on International Agricultural Research (CGIAR), as well as with some of the possible donor agencies.

ICIPE will continue its research to backstop the activities of the project through the available limited core funds. With additional funds, we the project could be extended to cover more than the original four countries in the region and include livestock ticks and tse-tse fly management strategies in the proposed network.

PART II

IMPROVING NORTH-SOUTH ENVIRONMENT CO-OPERATION

HOW WEST GERMANY REVAMPED ITS POLICY

by
Dr. Klaus Erbel
German Agency for Technical Co-operation

The German Government has recently reaffirmed that the conservation and rehabilitation of natural resources in developing countries is one of its priority aims in economic co-operation projects. A task force, created by the Federal Ministry for Economic Co-operation (BMZ), has developed over the last few years procedures, studies and other materials, e.g. checklists and matrices, to assess environmental impacts of projects , and to find and apply solutions for minimising detrimental effects to the ecology and to the human environment.

These procedures are to be used in appraising, planning, managing and evaluating all programmes, projects and measures having a significant environmental component, undertaken in the framework of German financial and technical co-operation. The philosophy behind the proposed environmental impact assessment procedures is that it must be considered as a more or less continuous process of scoping, replanning, monitoring and auditing, rather than in terms of a single environmental approval statement, valid once and for all.

We consider water to be one of the key elements in development, be it for agriculture, industry, infrastructure (water supply, energy) or human settlements. At the same time, water is subject to the highest environmental stress among natural resources because of:

-- Steadily rising demand, e.g. growing populations accompanied by growing awareness and demand for hygiene and the prevention of water-borne disease;

-- Decreasing quality, both in surface and ground water;

-- Decreasing availability (e.g. because of lowered water tables or dried up springs);

-- Irrational, i.e. unplanned, unbalanced, excessive or unco-ordinated use of water due to inadequate legislation, sanctions or planning instruments;

-- Climatic changes or human infringement on water resources through erosion, deforestation, overgrazing, etc.;

-- Salinity problems, e.g. due to innapropriate irrigation methods, insufficient drainage or overpumping of wells close to the seashore;

-- Environmental catastrophes like oil or chemical spills, or radioactive contamination.

German economic co-operation projects dealing with these problems can be divided into three types of measures:

-- To conserve natural resources;

-- To rectify or rehabilitate damaged ecosystems;

-- To strengthen institutions giving advice or training people to enable our partner organisations to help themselves.

In each case, the government of a developing country hands in an official project request which, in the case of technical co-operation, is then appraised by the Ministry for Economic Co-operation and by the Organisation for Technical Co-operation (GTZ) in terms of necessary inputs, time schedule and administrative procedures involving close consultations between support agency and recipient country.

In the context of this dialogue, it will be decided how the environmental problems might best be tackled, be it by:

-- Introducing new laws or modifying existing ones;

-- Developing procedures and control measures to enforce existing laws;

-- Strengthening organisations in charge of monitoring environmental parameters;

-- Doing studies on specific environmental problems;

-- Imparting technical, economic and organisational skills and know-how;

-- Providing advisors, instructors, experts and other specialists to partner organisations;

-- Supplying laboratory instruments and other equipment.

As these contributions will usually not lead to immediate productive outputs with positive cost-benefit ratios, they are normally granted on a non-repayable basis.

We believe that environmental issues have to be faced and solved by decision-makers and target groups in the developing countries themselves. Donors can help in the analysis of the situation. They may also give assistance in implementing demonstration projects. But for the bulk of the problems and in the long run, only initiatives by the local government and population count. More than in other fields of economic co-operation, external support agencies will have to restrict their activities to "Help for Self Help".

This means that particularly where basic human needs are concerned, as in the case of water resources, considerable information transfer, public awareness campaigns, and testing of the acceptance of the local population for certain measures to ensure they are able and willing to participate are necessary before implementation can start.

Water, unlike most other natural resources, is considered a common good, whose distribution affects the well-being of families, communities and peoples. One of the main problems in managing water resources in an environmentally balanced manner is to look after the interests and needs of all users. For example, in the case of a surface water reservoir, agreement and, where necessary, compensation has to be negotiated among:

-- Those living in the catchment area;

-- Those living on the shores of the reservoir;

-- Those downstream of the dam, who profit directly from the impoundment;

-- Those who want to use the water for drinking;

-- Those who want to use it primarily for irrigation or power production;

-- Those who may want to use the reservoir for sewage disposal; and

-- Those who try to preserve it as a special biotope or for fishing.

The broad interest in the use of water resources is reflected in the variety of ministries and other organisations responsible for their management. Sometimes the most difficult part of a project is to find the right implementation concept, to obtain co-ordinated approval of all concerned official bodies, and acceptance of the affected populations or non-governmental organisations. A rough check list of procedural steps to this end might look as follows:

-- Verify whether existing laws are sufficient to guarantee balanced use and development of water resources;

-- Ensure that these laws are enforced through monitoring and sanctions;

-- Develop national or regional water master plans describing resources, defining demands, fixing user priorities;

-- Create autonomous bodies, responsible for certain water resources, e.g. catchment areas;

-- Ensure that all users of this resources are adequately represented in such bodies;

-- Ensure that rights, duties, benefits, responsibilities and constraints are evenly distributed by way of a compensation system;

-- Monitor environmental changes, update master water plans; redefine utlisation of the resource if the ecological equilibirium suffers from over-exploitation;   and

-- Inform or train local people to enable them to understand why the water resource is managed in a certain way and how they can help to protect it.

WHAT FINANCIAL INSTITUTIONS CAN DO

by
Monique Barbut
Caisse Centrale de Coopération Economique, France

The history of industrialisation in the West over the last two centuries is full of environmental lessons that could be used to the advantage of the developing countries. However, it is not sufficient to adopt the principles, methods and criteria of the industrialised countries without taking into account the special characteristics of developing countries. That is why an economic approach to the "environmental" aspects of industrial projects must pay the closest attention to the valuation rules used.

The African economies are characterised, among other things, by a pre-eminent non-market sector surrounding the market economy. This gives rise notably to two consequences for economic evaluation. The first is that the possible effects of environmental change on the non-market sector are extremely difficult to estimate, with the frequent result that anything which is not quantifiable is ignored.

The second consequence is that, owing to the actual extent of the non-market sector, the market sector is not necessarily representative of the total economy and, in fact, cannot always provide the right information and signals for economic evaluations. This is another reason why it would be a mistake to resort to the criteria and parameters used in the industrialised countries.

Despite these difficulties, environmental concerns must be considered when evaluating projects in developing countries that involve major projects like use of underground or marine resources, industrial complexes, etc. Public opinion, especially in the donor countries, is very vigilant with regard to such projects. Unlike projects for public services, such as transport, communications or drainage infrastructures and networks aimed directly at meeting the needs of the population, industrial projects, and especially those launched by private firms, are usually perceived as profit-seeking, even though they, too, affect the commom weal (satisfaction of needs, job creation, tax revenues, external balance). Public opinion is therefore more demanding. Given this, pollution and risk prevention must be considered in any decision concerning industrial projects. The financial agencies involved in official development aid are no exception to the rule.

It is naturally up to the recipient governments to define the standards they want to see applied in their own countries. But public opinion in donor countries often attaches importance to ensuring that projects financed out of public funds are not harmful to the environment.

Consideration of Environmental Problems

Funding decisions are preceded by various studies (prefeasibility, feasibility) based on a range of investigations relating in particular to given technical characteristics of the project. More or less full consideration of measures for the prevention of pollution and industrial risks may lead to a number of alternatives whose financial implications can be assessed.

However, until now, environmental problems have all too often been approached through the industrial operators' use of what is regarded as the best available technology. The reasons for the choice made and the examination of possible alternatives must be opened up for a discussion of other technologies, experiences and achievements elsewhere and comparative costs. This can come as part of the various consultations held prior to taking the financial decision.

To this end, the terms of reference laid down by the investor concerning the various preparatory documents, studies and evaluations must explicitly incorporate this approach. The prefeasibility and feasibility studies must present different alternatives as early as possible, accompanied by an evaluation of the investment and operating costs. Similarly, the external factors (economic disadvantages) must be analysed, even if their evaluation is uncertain.

The financial agency must therefore obtain all the information available on these aspects in order to check the project's conformity with environmental constraints throughout its elaboration. However, this examination should not compound the difficulties by imposing excessive administrative constraints aimed at protecting the environment. Flexibility is also necessary to establish a relationship of trust between the different partners in order to find the best solution.

Initial Need for Information on the Project

The financial institution must request that the environmental aspects of a project be included right from the initial design stage. This request is made to the promoter of the project or to the consultants commissioned by the government charged with carrying out the feasibility study.

The very first documents provided on the project (generally the feasibility studies) should explicity discuss its environmental impact in a separate chapter, or if need be, in a special study.

The information needed to start an industrial project includes the following:

Main characteristics of the planned location:

-- Dominant winds if there is likely to be dust or poisonous smoke emissions;

-- Problems of water resources;

-- Flow and use of watercourses in the event of waste disposal into these watercourses;

-- Availability of raw materials;

-- Raw material flows;

-- Presence of populations and activities likely to be adversely affected by the project's location.

-- Description of the options according to a pre-established plan;

-- Water supply;

-- Manufacturing effluent;

-- Safety and accident risks and possible ways of reducing them for a specific site.

-- Description of the norms and technology used in industrialised countries and, where they exist, in the recipient countries;

-- Detailed description of the various possible alternatives;

-- Inclusion of environmental factors, and their weight in the comparison of different alternatives relating to major, not specifically environmental, options;

-- Adjustment of the demand for information according to the project.

Experience generally shows that systematic "environmental impact studies" mainly describing or studying the surroundings should be avoided as being too cumbersome or expensive. However, discussions of the procedures used show the advisability of collecting information as a function of the foreseeable adverse consequences for the environment.

A proviso must be made, however, for the following reasons:

-- The evaluation of environmental impact is pointless if it is known at the outset that it will be negligible;

-- The cost of this evaluation can be very punitive for a small project;

-- A project can change a great deal over the course of time, and it is not advisable to launch a full-scale study immediately when many options have not been decided or may be substantially modified.

The initial demand for information must therefore be qualified according to:

-- The size of the project;

-- The nature of the project (type of industrial or agricultural activity, etc.);

-- The number of options left open.

Checking the Information Supplied by the Proponent

In most cases, financial agencies do not have staff who are capable of checking the details of the technical information supplied by the proponent or the consultants responsible for the feasibility study and project design, nor is it advisable for them to burden their structures to do so.

Financial agency experts who examine projects play a fundamental role, and it is they who are responsible for consideration of the environment, with the possible help of outside partners (ministries, firms, universities, consultants).

Lastly, a dialogue on environmental matters is necessary within the financial agency among economic, financial and technical experts, as well as with the project's promoters and designers, the government and any experts consulted on special points.

It is thus important to prepare the staff of the financial agencies responsible for examining and evaluating the environmental component of projects. This may be done through:

-- Seminars (preparation, training);

-- A check list by type of project and by sector of industrial and perhaps agricultural activity.

Where the environmental implications are important, the financial agency may have to resort to a specialist consultant.

Follow-up of the Project and its Development

Various case studies show that a project can change significantly between the first feasibility study and its implementation. The environmental aspects therefore cannot be considered once and for all at one particular stage of the project: they must be examined at every stage, sufficiently upstream to avoid certain design errors and sufficiently downstream to enable the project to adjust to new conditions. Where choices must be made, comparisons among options can only be made on the basis of specific data regarding implications, costs, performance, arguments.

Terms of Reference for Tenders

At the request of the financial agency and depending on the information collected, the invitation to tender may include environmental obligations, such as, in the case of an industrial project, performance stipulations, information required, quantified alternatives, etc. In the absence of such an explicit request, environmental information may be left out of the terms of reference and hence not be considered in the tenders.

Place of Environmental Criteria in Decision-making

Individual responsibilities of the various parties involved

Consideration of the environment in a development project raises the question of the responsibility of the different parties involved: the financial agency cannot take the place of the government of the country where the project is to be carried out. In particular, it is not its function to fill the gap if there are no administrative procedures to examine the environmental impacts like those existing in the developed countries.

However, financial agencies providing development aid are not solely concerned with financial assistance to compensate for insufficient savings or foreign exchange. They are also concerned with the contribution projects make to "development", and may thus initiate development projects themselves or propose altering their components.

Compliance with environmental conservation , like the pursuit of other development objectives, may lead to contradictions among different objectives and hence to trade-offs between economic necessities and environmental or other non-economic factors.

The financial agency can intervene at different stages to improve consideration of the environment, for example, when decisions are called for during the elaboration and preparation of projects:

-- At different stages of the project's definition;

-- When tenders are being chosen;

-- When financial decisions are taken.

A financial agency providing development assistance, therefore, has many opportunities to ensure that the environment is properly considered during the projects it deals with. It is possible and advisable for environmental criteria to operate mainly upstream when the project is being defined (in the case of the chemical industry, for example, choice of water supply, location, ammonia supply).

As considerations of political expediency often enter into the financial decision, it is better if the main lines of the project are already defined with all due regard to the environment. However, differences may emerge between the environmental standards regarded as necessary by the financial agency and those adopted by the recipient government. It is conceivable that in the absence of what it regards as an acceptable solution, the financial agency might feel obliged to refuse its assistance if adequate preventive measures were not taken given important industrial safety or health hazards. But the generally low cost of prevention should make this type of situation unlikely.

The Cost of Environmental Considerations

It has often been thought that repairing damage to the environment was less expensive from a stricty financial viewpoint than prevention. The magnitude of certain recent industrial accidents has revealed the dangers and limitations of this view.

As in many other fields, the more thorough the consideration of the environment in prior feasibility studies, the lower the cost. This applies to study costs (collection of information, cost of expertise) when the project is designed and defined, thus avoiding the need to modify projects near or at their final definition.

It also applies to the project itself, for experience has shown that the extra costs associated with the late addition of an environmental protection system are very much greater than those incurred when specifications are included at the outset.

The financial agency can thus make a major contribution to effective and inexpensive consideration of the environment so that neither the design nor implementation of the project are affected.

ALIGNING PRACTICE WITH THEORY IN NIGER

by
A. Hamildil
Ministry of Agriculture and Environment, Niger
and
O. Hamil
Ministry of Co-operation, France

"Due regard to environmental considerations", "desertification control", "village land-use management" -- all these terms, while expressing specific concepts, in fact have much in common and reflect a new approach to rural development. Under the joint impetus of the Club du Sahel and the CILSS, all the Sahelian countries have, in principle, integrated these concepts in the framework of their respective desertification control plans. After a probing and comprehensive diagnosis of the situation, a new developmental approach is being proposed, giving rural communities a genuine say while at the same time favouring technical and multisectoral responses.

These studies implicitly point to the need for major reforms in such areas as:

-- Project content and design;

-- Implementation structures; and

-- Working methods of field officers.

Taking Niger as a case study and on the basis of the joint International Development Association (IDA)/French forestry project, an attempt is made here to define or identify one of several possible methodological approaches that could align practice with theory in achieving these reforms.

Under the auspices of the "Conseil National du Développement", the senior organ of the "Société du Développement", Niger has defined its policy through a process of in-depth discussions with all the parties concerned and helped to focus thinking in drawing up Niger's desertification control plan. The national charter, recently adopted by referendum, also incorporates the key factors identified in the discussions.

Methodological Objectives and Inherent Constraints

Any environmental approach must meet the following objectives:

-- Encourage rural communities to forge and manage their own development, with due regard to their environment and to evolving new, more
appropriate and higher-output production systems;

-- Structure and even standardise action, so that new projects can
build on the store of past experience;

-- Check that rehabilitation schemes and other investment projects are
replicable, sustainable and compatible with the country's financial
and human resources;

-- Respect the individual character of each project, while ensuring
that approaches are sufficiently harmonised, at least at the
district level, to ensure that projects do not vie with one another
but are mutually reinforcing;

-- Ensure that all the actors (central government, local communities,
etc.) enjoy the fruits of their endeavours.

But rural communities are subject to a number of major constraints that
point up the complexity of the undertaking. For one, villages are not
homogeneous entities. Within each village, farmers' concerns and behaviour
are determined by such factors as land tenure, soil quality, local customs,
and the like. This makes it difficult to achieve consensus on the action to
be taken. Moreover, villages are subject to such national economic controls
as grain prices, input subsidies, value of standing timber, etc. This limits
the possibilities for changing the production system, confining the population
to isolated initiatives that are all too often doomed to failure.

These problems are often compounded by a body of laws that have been
superimposed on traditional legislation. The lack of community involvement in
environmental management and the difficulties of reforming land tenure also
hinder the development of new approaches.

To overcome these constraints, a new approach is being taken based on
such general concepts as comprehensiveness (i.e. problems perceived and
resolved by way of a multidisciplinary and multisectoral approach); integration (action defined, implemented and managed by those directly concerned at
grass-roots level); and consistency, so as to ultimately establish a master
plan for the development of national land resources.

Linking Environmental Considerations With Other Key Development Issues

Integrating environmental and development concerns is a bit like a
balancing act: the environment is in a state of unstable equilibrium and one
teeters along on the way to development. The production system plays the role
of balancing pole; depending on how it is held, it will adapt to immediate
conditions, regulate and co-ordinate sectoral development with a view to
achieving overall equilibrium.

The quickening pace of population growth in Niger in recent years has
radically altered the country's socio-ecological equilibrium without, however,
changing traditional systems of production. Boosting production by bringing
more land under cultivation no longer seems feasible today.

Given these constraints, as well as the problem of drought, the only areas where some progress can conceivably be made are population policy (with its longer-term impact) and measures to change farming methods so as to increase the yield per unit of farmland.

To create a new balance, efforts must be directed at new legislative and land tenure reforms that better reflect the interrelationship between forestry, agriculture, animal husbandry and population. To achieve these aims requires a three-pronged approach encompassing development, training and research.

Proposals for the Implementation of a Development Strategy

Tackling the problems of land tenure is necessary to ensure that action is solidly based and sustainable.

As traditional rights have been whittled away by "modern" law, there is a real need to clarify and normalise land tenure rights so that individuals and communities can be sure of enjoying the fruits of their labour on the land they farm. This is a sine qua non for land-use planning.

In line with comprehensive, integrated "awareness-promoting" and the consistent approach advocated above, two priorities are proposed:

-- The first is community involvement in land use and land resource management. This encompasses all measures that either call for decision, action and compliance with community practice or else involve individual action with a long-term spin-off for the community.

As these problems obviously need to be viewed in a wider context than that of the village, it was decided to take the smallest territorial unit, namely the canton.

-- The second priority is individual action at the farm-plot level, in conjunction with co-operative action (voluntary joint farming arrangements), with a view to tackling the problems involved in changing production systems and increasing yields.

The problems to be tackled and the risks run by farmers are thus of a different nature than in the first priority and call for adequate extension services. For this reason it is best to work on the level of the village and the agricultural land within village limits.

In putting such a programme into practice, the first step is to select the villages that are the most representative, either for their agro-ecological mix or for similarities in configuration.

It must be remembered that in any attempt to systematise the approach, it is not possible to act on all fronts at once. But as a systematic approach implies that action should be countrywide, phased and orderly, this means that typical, large agro-ecological units should be pinpointed, in which co-operative development efforts geared to the two priorities set out above can initially be concentrated.

In this way, a first generation of pilot projects can be defined with a view to framing a new (rural and forest) land tenure code adapted to the needs and special features of each large administrative unit. These could serve as a valuable reference for second-generation projects.

For example, the Niger Forestry Project (IDA/FRANCE) suggests ranking the major agro-ecological units in the "département" of Niamey on the basis of cantonal boundaries. A representative canton has been selected in each zone. In this particular instance, the six representative cantons have been chosen and will be given priority with regard to first-generation projects.

Each local project to "change the production system" would give priority to upgrading technology in agronomy, water supply and crafts.

It is difficult to take a "global" approach to the implementation and management of cantonal projects without attacking the problems posed by integration. To solve problems of mobilisation, competition and even friction among the various technical services involved in the field to encourage village communities to take their own future in hand, responsibility for project management and co-ordination would be entrusted to a cantonal develop-ment bureau that would be part of the Société de Développement.

Even though each bureau could obviously not have a team of senior experts permanently on tap, it would be able to call, as and when needed, upon:

-- The skills of a multidisciplinary team made up of the most effective technical experts in the larger administrative unit;

-- The practical assistance of specialised consultants attached to the technical ministries;

-- A common methodological approach for on-the-spot implementation of field projects.

This approach entails a radical change in the working habits of field officers for it implies a collaborative effort in the framing of a development scenario. Once this scenario has been endorsed by all the parties concerned, it can be embodied in a contract that duly sets out mutual commitments.

Though no cash is handed out, it is important to encourage community participation. In recognition of their commitment to a programme of action, it is desirable that, at the request of the villagers themselves and after negotiation, the projects bring tangible benefits. Financing can be provided to improve infrastructure or the means of production and distribution (wells, bore holes, tracks, schools, clinics, seed, agricultural inputs, farm implements, educational and health supplies and vaccinations, among others).

Proposals for a Training Strategy

No development strategy can succeed without a training strategy, for it is the only way to ensure that action is sustained and that the drain on financial resources is limited. In the development strategy described here, training of managerial staff working in the local development bureau is aimed at developing multi-disciplinary extension workers rather than narrowly-qualified technicians. To administer this basic work force, the simplest solution would probably be to establish a field corps, attached to the regional offices.

As for senior officials, and in particular those who chair the sub-regional development boards and who would chair the management committee of the local board, their role is sufficiently crucial to warrant special efforts to enhance their awareness of environmental issues and environmental management expertise.

Training strategy should therefore be formulated in a regional context as well as nationally.

Efforts should be directed at introducing a training kit that deals with environmental concerns, so as to instil a comprehensive and integrated awareness of rural development and environmental management issues; framing a policy for further training, retraining and motivation of managerial staff (involving cash incentives); and grass-roots training.

The focus is in particular on:

-- Training school teachers who can play a crucial role not only at grass-roots level in educating young people for responsibility in the management of their own environment, but also in the village development council, in farm co-operatives and producer associations;

-- Training heads of cantons and village chiefs who play a key role in their respective development councils;

-- Training future administrators of co-operatives and voluntary producer associations, in co-operation with literacy centres;

-- Establishing craft training centres;

-- Helping to set up co-operative extension and self-help ventures at the level of the canton; bringing together farmers whose plots have been upgraded as a result of pilot projects and who could tell neighbouring villages about the new schemes.

A Strategy to Synergise Research Initiatives

The problem of research is an extremely complex one since the aim is at once to associate it with development with a view to more effective targeting and transfer, and to keep it just far enough removed from daily problems so researchers can lay the groundwork for the longer-term. Moreover, an appropriate research programme should take a regional approach and not forget national issues. This strategy is based on individual networks by topic or subject. Each network is responsible for programme co-ordination, follow-up and exchange of information.

These networks can focus the efforts of the decision-makers who determine the broad research thrusts and of the technicians who design the actual programmes. They also act as the spin-off point for the different structures responsible for managing, upgrading and applying research.

The networks are cross-country groupings based on a four-pronged approach: transparency, exchange of information, mutual assistance and on-going appraisal. This is sufficiently important to warrant that the establishment of networks and the appointment of the experts at their head be duly endorsed by nationally and internationally recognised bodies.

For Niger, these bodies are: the Consultative Group on International Agricultural Research (CGIAR); the "Conférence des responsables africains et français dans la recherche agronomique"; and a consultative group of national research institutes that might be represented in the Institut du Sahel.

In programme planning and design, the dynamism and discipline imparted by both the "Commission Consultative Nationale de la Recherche" and by the network boards or supervisory authorities will play a crucial role.

To sum up, integrating rural development policies with environmental considerations means that donor and recipient countries alike must be more rigorous in their selection and implementation of development assistance projects. However, adopting rigid planning procedures and approaches is out of the question, for no proposal, however attractive on paper, can ensure success. Flexibility is essential. Nonetheless, in order to safeguard the interests of the communities concerned, secure the government's credibility and gain from past experience, it seems necessary (at least at the level of the prefecture) to adopt some groundrules that help to determine the scope of each project. Compliance with such rules does not concern the recipient countries alone; in many instances, donor countries will also find themselves obliged to alter their approach.

PART III

THE ROLE OF ENVIRONMENTAL IMPACT ASSESSMENT

A STRATEGIC TOOL FOR SUSTAINABLE DEVELOPMENT

by
Frans W.R. Evers
Ministry for Housing, Physical Planning and the Environment
The Netherlands

Many of the environmental and developmental problems that confront us have their roots in sectoral fragmentation of responsibility for a highly interrelated set of problems. Sustainable development requires that such fragmentation be overcome. In the last few years we have learned that an integrated approach is a condition for achieving the desired environmental improvement. The need to integrate economic, social and ecological considerations in decision-making is a main element in a strategy for sustainable development. In order to be able to give form to such a policy, use has to be made of comprehensive environmental planning at all levels of decision-making. In situations where there is not yet such a sophisticated planning system, Environmental Impact Assessment (EIA) will be of use.

EIA can be viewed as a tool for both planning and decision-making. It contains two key elements: information and influence. EIA is concerned with identifying, predicting and evaluating the environmental effects of public and private development activities -- usually large scale projects such as dams, highways, industrial developments, etc. As such, it is primarily concerned with scientific methodologies and techniques (the information element). EIA is also concerned with procedures and processes for ensuring that the managers of both public and private enterprises consider the environmental consequences of the project and programme decisions they make (the influence element). In other words, EIA consists not only of the writing of a report through which information is provided to the decision-maker but also of provisions to make sure the decision-maker takes the information adequately and fully into consideration.

EIA has been identified as a major tool for the realisation of environmentally sound development. Over the last decade in many developed as well as developing countries EIA systems have been or are being implemented. In many cases EIA has brought about substantial benefits, including a reduction in the negative environmental effects of projects, the establishment of environmental values in government and industry decision-making and cost savings.

However, the use of EIA alone does not make a project sustainable: social and economic criteria have to be met too. Moreover, the sum of the projects that take environmental considerations into account does not automatically result in a pattern of sustainable development. Sustainable development planning is needed that makes use of the environmental, economic and social data that are collected both at the local level and at the national or even international level. However, sustainable development at the grass-roots level will, in the long run, only be successful if it is embedded in a national policy concept and supported by appropriate legislative procedures. EIA can play a role in providing the information that is needed to define the limitations that the environment imposes on development.

We are still a long way from a comprehensive system of sustainable development planning in developing countries as well as in industrialised countries. As long as a system of sustainable development planning is not in place the use of EIA will not only help to avoid adverse impacts of individual development projects, it will also make available the data that are essential for sustainable development planning. It is important that the local perceptions of ecosystems and the economic use that is made of them by the local population are included in the data.

The Work of the OECD in the Field of EIA

In December 1982, the Environment Committee included in its work programme a project on environmental assessment and development assistance. The project, reflecting also the interest shown by by the Development Assistance Committee, was carried out by an ad hoc group made up of officials from administrations having responsibility for development aid and for environmental protection.

At its first meeting the ad hoc Group on Environmental Assessment and Development Assistance agreed upon the following four objectives for the project:

-- To identify those types of development aid projects which, on the basis of past experience and future plans, are most in need of environmental assessment in order to ensure that due account is taken of the economic and social costs stemming from project induced damage to the ecology and resources on which the project is based;

-- To examine the constraints faced by developing countries in ensuring that environmental aspects are taken into account at an early stage in the planning of those development projects that are proposed for assistance; and, also, to determine how such constraints have in fact been overcome by certain countries either through their own efforts and resources or with external assistance;

-- To examine the experience of aid agencies in undertaking or failing to act effectively on assessments of the environmental aspects of development projects, including the nature and degree of influence of the assessments on the design, location, operation, cost or decision to abandon projects reviewed;

-- To determine, on the basis of the above and the results of similar work being undertaken by other international bodies, what types of procedures, processes, organisation and resources are needed to ensure that environmental assessment can be undertaken in a satisfactory, timely and cost-effective way, and to make proposals as appropriate to the Development Assistance and Environment Committees.

Environmental Assessment Approaches in OECD and Developing Countries

In examining the domestic environmental assessment requirements of OECD Member countries, the ad hoc group found that where environmental assessment has been introduced, a number of approaches have been taken which could be grouped into two categories -- formal-explicit and informal-implicit. The application of environmental assessment to development assistance activities is however much less advanced than for domestic developments.

To the extent that environmental assessment is carried out at all, it is, for the most part, done on an ad hoc basis. Only the United States aid agency (USAID) has a specific legal requirement for carrying out environmental impact assessments on certain types of activity.

Most aid agencies consider environmental factors in project/programme planning in a general way. Few, however, have specific procedures or guidance for routinely and/or systematically identifying those types of development assistance projects and programmes requiring an environmental assessment.

Although it is difficult to determine the exact number of developing countries which have initiated environmental assessment procedures on a regional basis, South-East Asian countries have shown the greatest interest, followed by Latin America, the Middle East and finally by Africa, where it is least developed.

The types of projects and programmes requiring assessment in developing countries generally correspond to those in OECD countries.

After reviewing the various approaches which have been taken to date in identifying projects and programmes in need of assessment, the ad hoc Group decided upon a list of activities which are generally agreed to have potentially significant impacts. The list incorporates those activities which have been assessed by aid agencies up to the present time. This list was incorporated in the Appendix to the OECD Recommendation on Environmental Assessment of Development Projects and Programmes [C(85)104] (annex 1).

This Recommendation also points out that, regardless of the project type, environmental assessment is always needed for proposed activities in certain fragile environments such as wetlands, mangrove swamps and coral reefs. It also lists a number of specific environmental effects which should be considered in carrying out an assessment.

Constraints to Carrying Out Environmental Assessment in Developing Countries

Generally speaking, the constraints identified in the various countries studied were:

-- Insufficient political awareness of the need for environmental assessment;

-- Insufficient public participation;

-- Inadequate or non-existent legislative frameworks;

-- Lack of an institutional base;

-- Insufficient skilled manpower;

-- Lack of scientific data and information;

-- Insufficient financial resources.

Aid Agencies' Experience with Environmental Assessments of Projects and Programmes

Nine governments presented a total of 16 case studies on environmental assessments which had been carried out by their aid agencies in Asian, African and Latin American countries. The case studies examined in particular the following topics: the form of the assessment; the way it was prepared; its content (including a description of the existing environment; alternatives considered; identification and assessment of environmental impacts; mitigation measures; and the effect of the assessment on project/programme decisions).

On the basis of the information in the case studies, the ad hoc Group identified five key elements for a successsful environmental assessment process. They are timing, personnel, "scoping", information and monitoring.

The first key element is timing. All the case studies pointed to the need to integrate environmental assessment at an early stage of the project planning. Where EIA is seen as an "add-on" or an "extra" to projects which have already been determined on the basis of their engineering, technical and economic feasability, it can perhaps suggest mitigation measures but can have no effect on project design.

The second important element is personnel. The success of an environmental assessment is very much dependent on the individual (or team of individuals) responsible for preparing it. Because of the large diversity of project and programme types to which assessment has been and can be applied, it is difficult to determine an ideal profile for an "environmental assessment preparer" which would fit every situation. Some types of projects can be adequately assessed by a single person with the right qualifications and experience working together with host government officials over a short period of time. Most projects, however, demand interdisciplinary teams of experts to carry out field investigations and data-gathering.

The third key element identified is "scoping". A crucial task in carrying out an environmental assessment is the determination, early in the project planning, of the most significant (i.e. serious) environmental impacts associated with a project and the reasonable alternatives available for executing the project in the most environmentally sound manner. Scoping is a procedure for accomplishing those tasks. An early meeting of the donor agency, host government officials, environmental experts and other interested parties to "scope" the assessment can result in quicker, less expensive and more efficient environmental assessments.

The availability of reliable scientific data and information is a fourth key element. It is important to work closely with local universities, research institutes and the affected public to obtain an insight into existing environmental conditions. As mentioned above it is important to guarantee that the local perceptions of ecosystems and the economic use that is made of them by the local population are taken into account.

Monitoring is the fifth key element the Group identified. One of the most important lessons to be learned from experience with environmental assessment is the need for monitoring the environmental impacts. Although it is not required by any aid agency, most if not all are coming to see the need for auditing completed projects not only as a sound management measure but also in order to test the accuracy of the assessments.

In its final report the Ad Hoc Group included the following statement regarding the relationship between EIA and sustainable development: "It should be stated that the general policy issue involved is how to promote economic development while preventing environmental degradation and protecting the long-term productivity of the natural resources on which development is based; in other words how sustainable development can be achieved. An environmental assessment process is one approach which aid agencies can take towards sustainable development."

A. A HYDROELECTRIC PROJECT IN INDONESIA

by
Philip Paradine
Canadian International Development Agency

The proposed Lake Sentani Hydroelectric Development is located in the Province of Irian Jaya in the extreme Northeast of Indonesia. The site is near Jayapura, some 20 km from the border with Papua, New Guinea. Sentani is a natural lake with an outflow through the Jafuri River eventually reaching the Pacific Ocean. By closing off the Jafuri outlet and diverting flow through a series of channels and tunnels to Yautefa Bay, it is possible to generate hydroelectricity.

The main environmental features of the project are therefore a reduction of the Jafuri flow, manipulation of the lake water levels, input of extra fresh water into the marine system of Yautefa Bay and terrain disturbances along the flow diversion corridor.

Sentani Lake is surrounded by 22 small villages whose residents live a traditional lifestyle. In the Yautefa Bay area fishing is practised while the diversion corridor is currently designated for high intensity development. The Sentani culture is very old and is traditionally orientated towards the lake with houses constructed on stilts in the water. This factor became a key aspect of the impact assessment.

When the Canadian-based consultant firm Acres International became involved in the Sentani Lake project in 1982, several studies had already been completed. In 1977 Tata consultants did a feasibility study which developed the flows through channels. The Tata study proposed a 10 megawatt plant and would have raised the lake level by two meters, requiring relocation of the Sentani people. Subsequently, a 1975 feasibility study was performed by NEDESCO and SMEC. That consortium, also funded by the Asian Development Bank, examined the Sentani proposal and redefined it to increase the installed capacity. As in the previous study, no environmental assessment was included in the terms of reference.

Form of the Assessment

In 1982 when the Canadian International Development Agency (CIDA) became involved in the project, Acres was asked to re-examine the situation. A proposal for a feasibility study and environmental reconnaissance was funded and work subsequently progressed to the design phase, during which a full environmental impact assessment was conducted. An interdisciplinary environmental team worked with the designers to develop alternative schema and integrate mitigation measures directly into the project proposal. The local population was directly consulted and the lifestyle of the lake-dwellers will continue to be possible after project development. The proposal submitted to decision-makers involved a 12-megawatt project and decisions are now pending on construction.

Content of the Assessment

There has been a growing awareness of the necessity to increase energy in this part of Indonesia. National policy is to promote development of some of the less populated areas of the country, thus creating the need for energy to supply the projected growth.

The projected population increase and associated industrial activity within the Jayapura region over the next four years requires an increase in the generation capacity of the existing electrical system.

The most economically attractive alternative involves use of the outflows from Lake Sentani as a potential hydroelectric supply. This would substitute for expensive diesel-generated electricity and permit supply to the provincial capital and surrounding region.

Scoping of studies of the existing environment

The environmental reconnaissance-level study involved the fielding of a team which was on-site for six to eight weeks. The team consisted of a civil engineer, a hydrologist, a scientist, an economist, and an energy systems planner. The team operated interdependently through daily meetings. Based on an understanding of hydroelectric projects in general, and a limited data base from previous studies, a generic list of impacts was established. The reconnaissance was intended to verify the data base and scope issues for further studies.

The team's first priority was to establish contacts with local government people, residents, the university, and anyone else who could provide baseline information and help identify the issues of importance for the local population.

During the environmental reconnaissance, the project concept became very important as it was obvious that the original two-meter water level increase would entail major impacts. The effect of alternative lake operating levels was determined as a critical area for further study and subsequent environmental information gathering was scoped accordingly.

Food availability for the Sentani people is directly related to the lake levels. Not only is the lake used for fishing, but the nearby marshes are harvested for sago. In addition, the shoreline behind the houses is used for the planting of vegetable gardens, so that lowering of lake levels during the farming season would be beneficial.

Farming was also potentially affected along the proposed corridor where rice and vegetable plots could be subject to disturbance by the construction of channels. Another aspect of the food availability issue concerned the tribes along Yautefa Bay, where traditional fishing lifestyles are followed. Possible disturbance by the sudden overflow of lake water into the marine environment was of concern.

An anticipated issue concerning lake water levels that was identified during the reconnaissance phase was that of public health. As the Sentani people use the lake as a latrine, too low a water level could spread disease, while too high a level would contaminate the shore.

Along the corridor local planning objectives had to be considered. Ridges and swamps alternate continuously along the route. Areas that are dry and suitable for housing are limited, and estrangement of prime potential housing lots was to be avoided. Lack of land registration was of concern and projected urban growth had to be taken into consideration. National planning objectives were also of importance, since a proposed transmigration area exists south of Sentani Lake and along the Jafuri. The impacts this project would have on the water supplies and land uses for the transmigrants had to be considered.

On the basis of issues identified in the field, a scope of work was prepared for the environmental assessment and alternatives proposed for operation lake levels.

Study of alternatives

During the preparation of the environmental assessment, local people were extremely important in providing site-specific information. Although the team spent four months working on the Sentani Lake, statistically significant information could not be collected in such a short time. Where necessary, the collective memory of the Sentani people was used in lieu. Information on historical lake levels, resource utilisation, and fishery in particular, was gained from the local population. Cultural information was an integral part of the data collection.

The major alternative examined during the environmental assessment was the Lake Level Rate Curve, which is the constraint governing manipulation of the water level during operation. A computer model of Sentani Lake was prepared to try various operating scenarios. The lowest constraint was dictated by sanitation levels while inundation of floorboards was the upper constraint.

For fishing, it is preferable to raise the lake level during the spring to allow the fish to spawn in their natural areas; with the water being kept high long enough for the fry to hatch and move down into the lake (thereafter drawing the lake down as fast as possible to allow people to plant their gardens).

Often a computer model is used to optimise the energy production without integrating environmental impacts. However, in this case, it was possible to achieve the same amount of energy while staying within all the environmental parameters.

As for the corridor alignment, values were placed on all of the relevant structures, so that every time a change was made, it was possible to recalculate how many houses, people, or hectares of crops would be impacted. Through land-use planning, it was possible to avoid cemeteries or schools and avoid cutting transportation paths. Hence, the corridor alignment was, in the end, quite different from that which had originally been proposed. In fact, the corridor was actually diverted considerably, to avoid land that was slated for future housing. It was also essential for the design to include safety considerations with regard to the canals.

As described above, real alternatives were considered in project design using information obtained during the environmental assessment. This resulted in some improvements being predicted for the project area including:

-- Improved fishing areas on Lake Sentani for people of Yoka near the approach channel;

-- More recreation areas or islands in the lake from generally lower lake levels throughout the year;

-- Improved sago stands and access to swamps around Lake Sentani from generally drier conditions resulting from lower lake levels;

-- Improved conditions in Yautefa Bay for milkfish and other estuarine species.

Environmental impacts

However, despite these improvements, a number of anticipated impacts remained. Potential mitigation measures and proposed compensation were therefore summarised in the environmental assessment as follows:

-- Construction of the hydro corridor and powerhouse will affect between 448 and 593 people, 67 to 97 structures, 30 to 45 ha of land (purchased or leased) and 25 to 145 ha of crops, depending upon the alternative chosen;

-- The village of Puay will be the most seriously affected community on Lake Sentani as a result of weir construction on the Jafuri River. Fish resources of the upper Jafuri River will be lost and a decline in fish catches on Lake Sentani near Puay is likely. The latter is expected due to anticipated water quality degradation in this end of the lake when the natural outflow to the Jafuri River is blocked by the weir. The drying up of the upper reaches of the Jafuri River will also adversely affect Puay's accessibility to agricultural lands and wildlife (and, hence, hunting) along the river;

-- The people of Lake Sentani will be inconvenienced by having to adapt their fish cages to a greater annual range in lake-level fluctuations;

-- The people dependent on Yautefa Bay fisheries are likely to be adversely affected in the short term by a reduction in fish harvests until marine resources in the bay adapt to an estuarine environment;

-- The virtual termination of lake flows to the Jafuri River will reduce mean annual flows in the Sunggrum River, although these are estimated still to be adequate for irrigation requirements currently forecast.

Mitigation measures

A number of proposals for mitigation were presented to alleviate impacts. These include:

-- The people currently occupying or owning lands along the hydrocorridor and at the Jafuri River weir site should be compensated for buildings, lands, and crops lost or damaged as a result of the project;

-- The people of Puay and Sekanto should receive sufficient compensation to enable them to continue their lifestyle with adequate fish resources and without the need to relocate their village;

-- Families with fish cages around Lake Sentani prior to operation of the hydro project should be provided with screening materials to eliminate potential problems associated with a greater annual range in lake level fluctuations;

-- Families who regularly fish Yautefa Bay as their primary resource base should be provided with additional gill nets to offset a possible reduction in fish harvests following construction and operation of the hydro project;

-- Footbridges with railings across the tailrace channels in the vicinity of the fishponds and at the tailrace outlet to Yautefa Bay should be constructed to allow people continued access to both sides of the channel in these areas;

-- Efforts should be made to restore the environment as far as possible following completion of the project, through grading, contouring, and planting.

Results of the Assessment

While the project has not yet been completed, the substantial changes in design without power loss already indicate the value of the environmental assessment.

Detailed actions have also been suggested for appropriate government agencies to take before construction of the project, including:

-- Ensuring no further development occurs on the corridor easement, including a 30-metre buffer zone on either side;

-- Providing careful inspection when staking out the easement in detail as discrepancies between technical drawings and site conditions can easily occur;

-- Organising preparation of compensation payments for building, land and crops, and ensuring consistent and equitable treatment of individuals (including compensation for lands temporarily disrupted during construction);

-- Planning for relocation of people and structures should begin well ahead of construction, and the affected populace should be included in the process. A reasonable schedule for relocation should be determined, and the people affected should be notified well in advance;

-- Keeping local residents well informed of project activities so they may adjust their own activities accordingly;

-- Appointing a responsible individual to manage compensation awards for the people living around Lake Sentani, the Jafuri River and Yautefa Bay who are expected to be impacted by the disruptions to fishery resources.

During construction, it is important that a safety and environmental inspector be employed as part of the construction management contract to ensure that all necessary safety precautions are in place and that environmental recommendations contained in this report pertaining to construction activities are followed. This should include supervision of daily reporting of fish catches in the villages of Puay and Sekanto and in villages around Yautefa Bay, since these data are vital to any subsequent monitoring programme following project implementation.

Monitoring following construction

Approximately three months following project operation, a survey of Puay, Sekanto, and Yautefa Bay fisheries should be undertaken to determine impacts and assess whether further mitigation is warranted. This should be undertaken by the safety and environmental inspector.

Approximately one to two years following commencement of hydro operations, it is strongly recommended that a more comprehensive environmental evaluation be carried out to compare post-project conditions with impact predictions.

The need for further mitigation based on the above assessment should be documented as part of the monitoring programme. Further environmental monitoring is recommended five years after hydro operation using a similar programme to that outlined above.

Constraints

While some government authorities in Jakarta were initially skeptical about the environmental assessment, attitudes changed as the results started to be known. It became evident that the project design and operation could change without affecting the cost-benefit ratio of the project. Government agencies, locally and in Jayapura, provided the study group with information requested, although constraints on horizontal co-ordination limited the ability of the team to discuss the project with various ministries.

Unfortunately, the university and environmental studies centre could not provide the group with technically skilled personnel. Although keen to assist in any way, the resources were not available to offer. Ultimately, however, a useful baseline study was completed. Fortunately, local people were always willing to give information about their lives and priorities and this compensated somewhat for lack of technical knowledge.

One specific constraint was a requirement that local people not be told about the proposed hydroelectric project. Therefore, questions had to be formulated in an odd manner, which tended to make them suspicious.

To sum up, the Sentani project demonstrates the value of incorporating environmental factors early in the planning process, such as during reconnais-sance. Because of this, key decisions were made early to allow changes to the project before designs were set. The value of scoping a list of issues to consider was also demonstrated. It allowed focussing on the right questions and eliminated costly and delaying studies.

Integration of the environmental team with the overall project team allowed a major impact potentially involving relocation of 6 000 people to be avoided. It also permitted optimisation of resource utilisation in the proposed project operations.

Because of effective communication with local people, the study was able to obtain information that was not available as published baseline data. This made a critical difference to project design. The involvement and co-operation of local agencies was also essential.

Finally, the importance of monitoring must be emphasised. It is impossible to quantify everything, especially with so little reference material on which to base some key predictions. Thus monitoring is absolutely necessary to make sure that the study and design are correctly verified and implemented, and that the mitigation measures that were proposed actually work.

B. THE GREATER CAIRO WASTEWATER PROJECT

by
Mohamed Talaat Abu-Saada,
Cairo Wastewater Organisation, Egypt
and
Stephen F. Lintner,
U.S. Agency for International Development

The Greater Cairo Wastewater Project, undertaken jointly by the Arab Republic of Egypt and the United States, is an excellent example of how environmental assessments can be used to assist both host countries and donor organisations in the evaluation of phased implementation strategies for major projects, the selection of technology, the evaluation of operation and maintenance issues, and in the identification of complementary projects to assure sustainable performance of the project. This experience can be helpful when preparing environmental assessments for other projects in developing countries.

In 1976, the Arab Republic of Egypt embarked on a massive undertaking to improve the wastewater collection, treatment and disposal systems for the capital city of Cairo. The objectives of the project were to improve wastewater collection, conveyance, treatment and disposal in the metropolitan region. The implementation of this undertaking has been assigned to the Cairo Wastewater Organisation (CWO), while the operation of the system is the responsibility of the Cairo General Organisation for Sewerage and Sanitary Drainage (CGOSD). The project involves not only extensive rehabilitation of existing facilities, relief pump stations and force mains, but new construction as well, including major wastewater treatment plants and disposal systems. Additional project activities will provide extensive support for institutional development and training programmes. In addition, the Ministry of Health will conduct water quality monitoring on a regular basis.

The city is geographically divided into two banks, East and West, by the River Nile. The East Bank, the oldest portion of the city, has an extensive wastewater system dating back to 1906. The West Bank wastewater system, which was constructed in the 1930s, is not as extensive, and as a result, the West Bank has a higher proportion of its population living in unsewered areas.

During the planning stages for the Greater Cairo Wastewater Project, consideration was given to these various structural differences and the varying needs of the areas. As a result, East Bank improvements focus on the construction of a major conveyance system, to carry existing and future wastewater to a new treatment plant at Gabel el Asfar. Construction activities for the West Bank are more extensive. They include the expansion of basic collection and conveyance facilities in the extensively unsewered areas, rehabilitation and expansion of the Zenein wastewater treatment plant, and the construction of a greatly expanded wastewater treatment plant at Abu Rawash.

The Need for the Project

Cairo, like many capital cities in the developing world, has been faced with the problem of supplying wastewater services to a population which is growing as the result of both rapid natural population growth and high rates of rural-urban migration. The present population of the metropolitan region is estimated at approximately 8 million and is projected to reach approximately 13.6 million by the year 2000. In serving a population that is increasing at a rapid rate, the Cairo wastewater system, which was originally designed for a population of less than 1 million, had become seriously overloaded and deteriorating to a point where unsanitary conditions were developing throughout much of the city. Inadequate investment in maintenance, especially for pump stations and sewers, further reduced the efficiency of the system. In addition, many areas around Cairo have never been sewered, even though many had been supplied with piped water.

Presently, about 66 per cent of the population is served by the existing sewerage system and 34 per cent reside in unsewered areas. The reliability of collection in the sewered areas is reduced due to the general overloading of the system and inadequate pumping capacity. In the unsewered areas, populations rely on a combination of public and private services for the collection of wastewater from vaults below or adjacent to houses. These devices frequently overflow or become inoperable, resulting in the large-scale flooding, ponding, and pollution of entire neighbourhoods. The high cost of commercial collection of wastewater in the predominantly lower income unsewered areas discourages all but the most essential use of the sewage disposal pits in these areas and promotes a variety of improper disposal practices. Because of these conditions, studies have shown that the rate of illness is higher in the unsewered urban areas of Cairo than in most rural areas in Egypt.

Approximately one half of the sewage collected receives partial wastewater treatment prior to disposal. The remainder, including that collected from unsewered areas, is disposed directly into open drains originally constructed for agricultural purposes which have become an element of the urban wastewater infrastructure. These drains eventually discharge into the Nile delta resulting in local degradation of water quality. Wastewater from industry and thermal power generation is not a significant problem in Cairo due to the concentration of these facilities to the north and south of the city outside the service area of the system.

The negative impact of this situation on environmental health and water quality is recognised by the Government of Egypt which has given top priority among infrastructure investments to the Greater Cairo Wastewater Project. The project has broad recognition of the need to make improvements in the system. This attitude has been important as there has been considerable local disruption of traffic and business during the implementation of the project.

Donor Support

The Government of the United States acting through the United States Agency for International Development (USAID) and the Government of the United Kingdom acting through the Overseas Development Administration (ODA) have provided capital and technical support for this project, as have private British banks. Assistance is also being provided by the Federal Republic of Germany and Japan. Local currency costs are being provided in part by the Government of Egypt. A major element of technical assistance has been support for the design of both the rehabilitation and new construction phases of the project by a jointly financed British-American engineering consortium named AMBRIC. AMBRIC works in collaboration with a consortium of Egyptian firms.

The total project cost is expected to be more than $3 billion. Supplemental studies, including Environmental Assessment, Tariff Studies, Unsewered Area Studies, Environmental Health Review, etc., cost $1.5 million and were funded by USAID.

Rehabilitation Phase

The existing wastewater system includes over 400 km of common sewers, 82 pneumatic ejector stations, 95 conventional pumping stations, and approximately 120 km of major collectors, in addition to five wastewater treatment plants. Due to excessive deterioration of the system, a multi-faceted rehabilitation programme was implemented between 1980 and 1986.

The work supported under the rehabilitation phase of the project consisted of major and minor repairs; structural and equipment modifications; debris and grit removal; and general cleanup of the system. Work was conducted on five major system elements: collectors and sewers, ejectors and ejector stations, pumps and pump stations, force mains, and treatment plants. This phase provided for a substantial improvement in the performance of the existing system.

The rehabilitation phase not only reduced the problems associated with the wastewater system, i.e. flooding and ponding of sewerage, but it also provided a foundation for developing most of the detailed plans and construction specifications for expanding and improving the system to serve the population of Greater Cairo.

New Construction Phase

On the East Bank, a major conveyance system is planned which will carry wastewater from the existing sewer collection system to a major tunnel pumping station at Ameria, which will have installed a centrifugal pump. From Ameria, a new 15 km culvert conveyance system will transport wastewater northward to two plants: a new locally designed treatment plant at Shoubra el Kheima and a major new activated sludge treatment plant at Gabal el Asfar. The Gabal el Asfar wastewater treatment plant will be a non-nitrifying activated sludge plant with thickening and drying facilities.

Construction activities on the West Bank have been designed to improve and expand the wastewater system and to assure its proper management. To meet these design criteria, construction will include: deep collectors, which will allow the elimination of 12 existing pumping stations; steep gradients to assure an adequate scouring velocity where incursion of sand is a problem; and simple archimedean screw-type pumps for major pump stations. With regard to the expandability of the system, primary collectors will have the capacity not only for current volume, but for extension of services to present unsewered areas and to adjacent developing areas.

Besides providing sewerage services to unsewered areas, additional activities focus on construction of a new treatment plant at Abu Rawash. The plant is a non-nitrifying activated sludge design and is expected to treat flows reaching 400 000 cubic meters per day. The design minimizes both energy cost and maintenance. Based on the examination of the system, the Zenein plant was the only treatment plant considered suitable for retention in the system. The plant will undergo extensive modifications, during this phase, which will result in an operational capacity of 300 000 cubic meters per day.

Operational Assistance and Training

In order to assure the reliable performance of the Cairo wastewater system the Government of Egypt and USAID have started implementation of a series of institutional development and training programmes in operation and maintenance. The programme has focussed on the development of improved institutional capabilities in administration, planning and financial management. Extensive support has been provided for the "training of trainers" in a wide variety of professional, administrative and technical skill areas. Special attention has been given to critical problems such as the management of grit accumulations in the sewers, sewer cleaning, pump station operation, and wastewater treatment plant operations.

The West Bank Environmental Assessment

USAID is required by United States law (22 CFR 216, "USAID Environmental Procedures") to prepare environmental assessments for all projects which are anticipated to have a potentially significant impact on the environment. All major water and wastewater projects are specifically required under this legislation to have an environmental assessment prepared to ensure that they are planned, designed and implemented in an environmentally sound manner. The Arab Republic of Egypt, although not specifically requiring the preparation of environmental assessments, requires that proposed projects be reviewed for compliance with a variety of laws and regulations concerning the environment.

Under USAID regulations, an environmental assessment is defined as a detailed study of the reasonably forseeable significant effects, both beneficial and adverse, of a proposed action on the environment. The objective of an environmental assessment is to identify potential environmental consequences of a proposed project to ensure that the responsible decision-makers in both the host country and USAID make an environmentally informed decision when reviewing and approving a proposed project and implementation plan. Included in the assessment is a detailed evaluation of alternatives to the proposed project and the identification of mitigation actions which might be adopted to eliminate or reduce unavoidable negative environmental impacts.

It should be understood that under the USAID approach environmental assessments do not recommend a specific course of action, nor do they determine whether a project should or should not be undertaken. These decisions are reserved for resolution by USAID and host government personnel during the process of project design, review, approval and implementation. This is important, as it makes the assessment a "dynamic" tool to ensure environmental soundness, rather than a "completed" document prepared to assure compliance with a regulatory requirement. The value of an environmental assessment in the USAID system is that it provides information concerning key environmental issues, an analysis of alternatives and reviews potential mitigation actions. This information is then evaluated with other detailed analyses relating to engineering, economics, management, training, and financing to provide an effective and environmentally sound project design and implementation plan.

An agreement was reached early in the design of the Greater Cairo Wastewater Project that the Government of the United States would provide assistance for capital construction on the West Bank of the Nile and the Government of the United Kingdom would provide assistance for capital construction on the east Bank of the Nile. Initially, it was anticipated that the AMBRIC model used for design of the system could be extended to joint USAID-Overseas Development Administration preparation of a detailed environmental assessment. However, for a number of reasons this did not prove possible and USAID proceeded to fund the preparation of a detailed environmental assessment for the proposed West Bank construction programme.

Preparation of the Environmental Assessment

The Cairo Wastewater Organisation and USAID recognised the need for the preparation of a detailed environmental assessment from the earliest stages of project development. An element of the initial design studies prepared by the AMBRIC Consortium included a preliminary environmental review of the project. The Washington based Environmental Coordinator of the Bureau for Asia and Near East of AID made preliminary site visits to the project area and held discussions with representatives of CWO during April 1979 to review environmental issues associated with the proposed rehabilitation phase of the programme and in March 1980 to review the proposed new construction programme on both the East and West Banks of the Nile. The scope of work for the Environmental Assessment was prepared by the Environmental Coordinator with the assistance of CWO during visits to Egypt during November 1980 and March 1981. The Environmental Coordinator returned to Egypt to supervise the initial phases of field data collection with representatives of the consulting firm retained to prepare the assessment and in October 1981 to participate in the CWO sponsored "scoping session". The planning visits to Egypt allowed for the advance collection of a variety of data and for co-ordination with the AMBRIC and Egyptian consortiums.

The environmental assessment was prepared for the General Organisation for Sewerage and Sanitary Drainage (GOSSD) and the Organisation for Execution of the Greater Cairo Wastewater Project (CWO) by American and Egyptian consultants. The total cost for preparation of the assessment was approximately $270 000 which was grant-financed by USAID. It was prepared over a 12-month period which included significant periods for the review of draft versions of the document.

The study was prepared by a 12-person inter-disciplinary team of experts from the United States and Egypt which included specialists in: agricultural engineering, agronomy, economic analysis of capital projects, economic analysis of natural resources issues, environmental engineering, Egyptian law, industrial pollution control, public health, soil science, social science and wastewater systems operations and maintenance. The assessment was prepared in two volumes: an Executive Summary (in Arabic and English) and a Main Report (in English with an Arabic summary and table of contents).

Preparation of the environmental assessment for the Greater Cairo Wastewater Project included the conduct of the first environmental "scoping session" held in Egypt. A requirement under USAID environmental regulations, a "scoping session" is a meeting of knowledgable and potentially affected parties to review the proposed scope of work of the environmental assessment and to provide advice concerning the preparation of the study.

The session for the project was hosted and chaired by the Cairo Wastewater Organisation and had 31 participants. These included representatives of Ain Shams University, AMBRIC, Ministry of Agriculture, Cairo Wastewater Organisation, General Organisation for Physical Planning, Ministry of Health, Ministry of Irrigation, National Committee on Environment, University of Alexandria, and USAID. The preparation of the assessment benefitted significantly from this session which allowed for the improved targeting of field efforts, identification of key sources of data, and the establishment of high level contacts with senior representatives of major governmental and technical organisations.

Major Issues Reviewed in the Environmental Assessment

The environmental assessment focussed on the review of the current environmental conditions in the greater project area, an analysis of the causes of these problems, an analysis of alternative wastewater management plans, a review of wastewater collection and conveyance alternatives and their environmental impacts, and a review of wastewater treatment and disposal alternatives and their environmental aspects of the following issues:

-- Alternatives for the sequence of facilities construction;

-- Alternatives for wastewater treatment; and

-- Alternatives for effluent disposal.

Each of the alternatives was reviewed with regard to its cost, its reliability under local conditions, the associated environmental health benefits, and its institutional requirements and social acceptability in the Egyptian context. The assessment emphasized the evaluation of alternatives with regard to both the impact of new facilities as elements of a well-conceived and well-run system, but also those impacts which could result if portions of the system do not function as intended. It also analysed constraints to efficient operation such as inadequate tariffs, non-enforcement of sewer use ordinances, and inadequate resources for spare parts.

The project design of the Government of Egypt and USAID made extensive use of the environmental assessment in the development of a strategy for phased investment. Based on the assessment, first priority was given to collection and conveyance investments, with second priority given to treatment disposal investments. The environmental assessment also was used to justify the need to identify and obligate significant additional funding by both governments to assure successful implementation of the complete project.

It was recognised that the emphasis on collection and conveyance would provide for rapid and significant improvements in environmental health for large numbers of residents in areas which were either unsewered or subject to routine flooding due to inadequate conveyance. However, it was understood that this decision would continue, on an interim basis, the long standing practice of discharging untreated wastewater to agricultural drains.

It should be noted that the analysis included in the environmental assessment showed that while this would temporarily result in a minor negative incremental impact to water quality, the benefits obtained from construction of permanent facilities for the removal of untreated wastewater from densely populated areas justified this decision. In addition, the risk associated with this investment strategy was limited due to the small amounts and restricted range of industrial pollutants discharged into the West Bank collection system.

Lessons Learned About Environmental Assessments

Timing is all important

The experience of the Greater Cairo Wastewater Project demonstrates that environmental assessments for major capital development projects can be prepared in a cost-effective and timely manner when they are prepared at an appropriate point in the course of project development. The Cairo Wastewater organisation and USAID believe that the funds spent to prepare the environmental assessment represents an effective expenditure of $270 000 to support the approximately $1.4 billion of new construction for the West Bank section of the Greater Cairo Wastewater Project.

USAID experience is that environmental assessments of major capital projects are best done following the preparation of the preliminary feasibility study, which allows for a clear identification of the proposed project and alternatives. The environmental assessment, to be an effective tool in decision making, should be available for concurrent review with the feasibility study. USAID does not recommend that any project be authorised to go to final engineering design and/or construction prior to the preparation, review and clearance of a detailed environmental assessment.

The costs of environmental assessments can be minimised if selected data collection needs are identified early and are included in the basic data collection programme for the engineering feasibility study. Savings can also be achieved by requiring the engineering consultant to allow the environmental assessment team to use base maps, system plans, technical data and cartographic drafting bases.

EIA should be an on-going review process

The experience of the Greater Cairo Wastewater Project illustrates that the initial environmental assessment is only one element of a continuous environmental review process which should be used by host countries and donors in the planning, design, implementation and operation of a major capital development project.

It should be noted that the Greater Cairo Wastewater Project was actually subject to a series of field-based environmental reviews by USAID environmental personnel: a) prior to the preparation of the detailed environmental assessment, b) during the technical review of final engineering designs for selected elements of the project, and c) through periodic field reviews.

Most important in assuring sound project implementation are the annual reviews of the water and wastewater sector which are held by the Governments of Egypt and the United States. These sessions allow for the routine assessment of progress, the timely identification of problems, and the joint resolution of issues. This continuous process is critical to assure that a project is planned, designed and implemented in an environmentally sound fashion as opposed to only being subject to the preparation of a detailed environmental assessment as a legal or policy requirement.

EIA can help establish phased investment strategies and technology selection

The experience of the Greater Cairo Wastewater Project demonstrates that environmental assessments can be designed to provide insight into complex project design decisions such as the selection of phased investment priorities and technology selection in a large-scale project. The environmental assessment was an important tool for the Cairo Wastewater organisation and USAID in making the difficult decisions on the sequence of construction elements in order to optimise environmental health benefits in a project with an implementation period of over a decade. The environmental assessment assisted in the selection of technology for wastewater collection, treatment, conveyance, temporary disposal and permanent disposal.

The usefulness of the environmental assessment in the project design and implementation process was enhanced by the fact that the analysis of all technical alternatives and mitigations proposed in the environmental assessment included an evaluation of their capital cost (foreign currency, local currency), recurrent cost (foreign and local currency), institutional development and training requirements, and the identification of the responsible implementing organisation. This information proved critical to decision making as it provided information on the cost and managerial implications of various options.

EIA should provide an integrated analysis for programme planning

The experience of the Greater Cairo Wastewater Project demonstrates that an environmental assessment can influence the decisions of host country and donor officials identifying major problems requiring resolution, providing a basis for policy, review, and serving as advocacy documents to obtain

114

support for project funding. The environmental assessment identified a series of issues which represented generic problems in the management of wastewater in Egypt. It reviewed the problems of institutional development, operation and maintenance, system reliability, and the financing of recurrent costs. By providing an integrated and objective analysis of the current status of wastewater services on the West Bank of Cairo and an assessment of the potential human health impacts of this situation, the assessment served as an advocacy document for justifying the provision of support for complementary project activities in institutional development, operation and maintenance, and training.

A major outgrowth of the assessment is the Water and Wastewater Institutional Development Project approved in 1985 with joint Government of Egypt and USAID funding of $20 million. The assessment brought to the attention of the host government and donor organisations the serious negative environmental impacts which can occur when systems fail to operate as planned due to inadequate design, poor construction or improper operation. It stressed the critical role played by properly prepared institutions and trained personnel in assuring that the environmental objectives of the project are achieved on a sustainable basis.

The importance of "scoping sessions"

The experience of the Greater Cairo Wastewater Project demonstrates the utility of conducting "scoping sessions" as part of the process for the preparation of environmental assessments. The USAID Environmental Procedures require that "scoping sessions" be conducted as an element of the environmental assessment preparation process. However, the experience in the Cairo study, and others, indicates that the sessions provide an important mechanism to assure widespread knowledge of the proposed project, the potential environmental impacts, alternatives, and possible mitigation activities.

It provides a forum for the participants, the project sponsor and the assessment team to interact to obtain a consensus on such things as the critical environmental issues which should be reviewed, the critical organisations and individuals which should be contacted, and location of the important sources of data. It also provides a means for establishing contacts to ensure that senior level personnel instruct their staffs to provide assistance to the team and logistical support for field visits. The Cairo Wastewater organisation and USAID attribute much of the $78 000 in savings which was realised by the consultant in the preparation of the assessment as attributable to the contacts made at the scoping session.

The advantages of a "joint team" approach

The experience of the Greater Cairo Wastewater Project demonstrates that the use of joint teams comprised of personnel from the host country and international consulting organisations is a technically sound and cost-effective practice. The preparation of the environmental assessment benefitted from the use of an Egyptian private sector consulting organisation as a subcontractor which provided both professional personnel, support personnel and logistical assistance. This association increased the efficiency of the personnel provided by the international consultant and provided direct access into the well developed Egyptian consulting community. It also reduced the time required by Cairo Wastewater Organisation and USAID to support the field operations of the international consultant.

The "joint team" approach provided the Egyptian subcontractor with an opportunity to expand their area of expertise and to develop a potentially long-term relationship with a firm from the United States. USAID has continued to use this approach to the preparation of environmental assessments throughout the Asia and Near East Region. For example, a major Government of Pakistan-USAID environmental assessment was prepared recently by a joint team which included five experts from an American consulting firm and five experts from an associated Pakistani firm.

Host governments and donors should recognise, however, that the use of joint teams requires the adoption of a policy which promotes collaborative preparation of consultant studies. USAID believes that the joint preparation of environmental assessments is an effective technique for the transfer of this methodology when this is a planned objective of technical assistance and provisions are made in the consultant contract to assure this will occur. It is recommended that when joint teams are proposed to prepare environmental assessments the scope of work should include a provision for the international firm to review the objectives and methodology of the environmental assessment with the local firm. The scope of work for the local firm should include provision for review of local customs, laws, regulations and institutions with the international firm.

The need for a flexible review and comment process

The experience of the Cairo Wastewater Project demonstrated the need to adopt a flexible approach to the review and comment process for environmental assessments. In most countries, developed and developing alike, personnel in governmental and non-governmental organisations are limited in their ability to review and comment on the large amounts of material which are routinely submitted to their offices. The traditional system used in the United States of soliciting formal written comments in response to a draft assessment is not an efficient way to obtain comments in the Near Eastern context. For a number of reasons, it is difficult for many organisations to provide written comments in a timely fashion (60 to 90 day review and comment period). After a major extension of the review period, the Cairo Wastewater organisation and USAID conducted visits to a number of key individuals to obtain their comments. This proved to be a satisfactory although informal means of obtaining responses to the draft and final assessment.

As the result of this type of experience in Egypt and other countries in the Asia and Near East Region, USAID has adopted a mixed approach to review and comment on assessments which includes formal written comments, small group meetings to review draft documents, and consultations with key individuals. USAID has found that the preparation and distribution of independently bound executive summaries greatly assists in providing senior level personnel, who do not have the time to review the complete assessment, with an opportunity to review the major findings and recommendations of the study.

EIA TRAINING FOR DEVELOPING COUNTRIES

by
Brian Clark
Centre for Environmental Management and Planning
University of Aberdeen

A principal concern of OECD is to consider how aid and environmental agencies of Member states might develop training initiatives in environmental assessment in developing countries. The OECD Council recommended measures to facilitate the environmental assessment of development assistance projects and programmes [C(86)26 (Final)]. In Annex II, it is stated that:

"OECD Member countries' aid and environmental agencies could institute training courses in environmental assessment. The provision of training should be made to a number of target groups in the host countries including elected representatives and senior decision makers in government and business, high level administrators, project managers, technical specialists, members of review bodies and representatives of environmental interest groups. The specific type of training to be undertaken would vary depending on the target group. For policy makers, for example, seminars should be conducted to demonstrate the negative effects which result from a failure to incorporate environmental elements in economic development planning and emphasize the benefits to be gained from environmentally sound planning. Training for project managers and technical specialists would emphasize procedures and methods for environmental assessment and their role and significance in environmental management."

We shall first consider certain general requirements for environmental assessment training before suggesting how donor agencies might positively contribute. This approach is adopted because much training in developing countries appears to be unco-ordinated and unstructured, a situation that is also found in a number of developed countries. Training programmes offered reflect a number of contradictions: on the one hand, idealism and intellectual committment; on the other, a lack of suitable manpower and financial resources, opportunism, entrepreneurial ability (to attract financial support) and market factors.

The relationship between supply and demand is often unknown, and there appears to be a lack of systematic awareness of what training does exist and more importantly whether it is appropriate. Training as a priority is now being given great emphasis by many international, national and local agencies, but it appears that there is often a lack of co-operation and collaboration in many of the initiatives taken.

The views on training requirements expressed here are based on more than ten year's experience of training over 2 000 participants from both developing and developed countries; negotiating with sponsoring organisations concerning programme content and finance; and conducting environmental assessments in developed and developing countries. Wherever possible, the points that are made are substantiated, but one of the real problems is the dearth of knowledge that exists relating to training in environmental assessment both qualitatively and quantitatively in developing countries, a comment that also applies to many developed countries.

Constraints on Training

There are constraints that have a direct effect on the ability of developing countries to implement sound environmental management and assessment strategies. With regard to training, a number of general and specific points must be highlighted:

-- There is no general and universal definition of environmental assessment. While there may be a general consensus of opinion of what it seeks to achieve, i.e. to predict the environmental effects of adopting a particular plan, programme or project, the mechanisms to achieve this vary from country to country. For this reason making general recommendations on training requirements may have hidden dangers, given the differences that exist in institutional arrangements, legal and procedural requirements, and levels of scientific knowledge when conducting environmental assessments;

-- Environmental assessment training can be considered as an activity in its own right or as an integral part of training for environmental management. It is also necessary to consider whether training in environmental assessment should be linked to an understanding of other evaluation techniques such as risk assessment, cost-benefit analysis and planning evaluation techniques (planning balance sheets, goals, achievement matrices);

-- Should training in environmental assessment lead to the establishment of a new professional group? In many developing countries, most of the people concerned with environmental assessment have other professional duties. It would therefore seem wrong to develop training to establish a new, professional EIA cadre;

-- Training initiatives in environmental assessment could be undermined if those who develop expertise are not able to satisfactorily perform their duties. Two specific areas of concern should be noted. First, there is a need for sufficient scientific knowledge, which will include an adequate data base so that those producing, and those reviewing, environmental assessments can adequately deal with documents such as Environmental Impact Statements (EIS). Secondly, a legal/procedural framework for co-ordinating environmental assessment will be required which as well as appearing sound on paper, can be properly implemented. It is clear that many of the

118

difficulties and much of the frustration expressed by those working on environmental assessment in developing countries relates to these two topics;

-- As a predictive tool and procedural mechanism it can be argued that environmental assessment requires a multi-disciplinary approach. Though lip service is paid in many developed and developing countries to the need for such an approach, it is difficult to achieve for many reasons. These include institutional conflicts between government departments and the strict demarcations between faculties and departments that exist in many universities and training institutes.

General Training: Needs and Demands in a Climate of Change

It is impossible to state with any degree of accuracy current and future demands for training in environmental assessment. This applies to developed countries but in particular to developing countries. It is equally difficult to indicate the current supply of training. The only figures that exist relate to general training on environmental topics, and the accuracy and relevance to environmental assessment training are problematic.

The question of need is also extremely difficult to forecast. A study by the East West Environment and Policy Institute suggests that approximately 10 000 professionals in tropical developing countries must acquire environmental skills within the next ten years. The percentage required for environmental assessment is not stated. In the European Economic Community, a Directive on environmental assessment became mandatory in July 1988, and it was estimated that approximately 2 000 people (decision makers, EIA project managers, technical specialists, review body members and environmental interest groups) trained in various aspects of environmental assessment would be necessary to implement it.

There can be no doubt that there is a real need for training in many aspects of environmental assessment and that this need is greatest in developing countries. There is also evidence to suggest that the atmosphere both in a political and economic sense may now be more conducive to meeting current needs for training in environmental assessment.

At the international level, recommendations by the Brundtland Commission, the World Industry Conference on Environmental Management, policy statements by the President of the World Bank and the proposals of a WHO working party on the Health and Safety Component of EIA, all explicitly state that greater resources should be allocated to fostering environmental awareness, evaluating the environmental components of proposed projects and environmental policy formulation. It is imperative that training should be an integral part of such initiatives.

There is evidence that in many developed countries now entering the post-industrial society phase, one of the greatest areas of growth, measured both in investment and employment terms, is in anti-pollution activities and related environmental control technologies. This means that there is now an increasing pool of manpower available in developed countries with the potential to provide specialised inputs to the environmental assessment process in developing countries.

Training in Developing Countries

Given the need to develop training initiatives in environmental assessment in developing countries, three closely related issues must be considered:

-- The target groups who would benefit from training;

-- The types and levels of training required to achieve the many objectives of an environmental assessment process; and

-- General principles concering the content of training programmes.

Target groups

Although the approach to EIA varies from one developing country to another, it is possible to identify certain distinct groups that will normally be involved:

-- Decision-makers/control authorities. This group comprises those who control and operate an EIA system in a country. Normally they will have power to indicate which projects (whether proposed by aid agencies, multi-national developers or public and private developers within a country) should be subject to an environmental assessment (screening) and what the study should cover (scoping). They comprise those who may legislate environmental assessment processes (political decision makers), those authorised to implement EIA by laws and statutes, and provide technical advice and assistance (government administration in central and local government) as well as review authorities in countries where this exists or is planned for the future;

-- Development proponents. This group comprises a wide and diverse range of those responsible for plans, programmes and projects which may be subject to environmental assessment. Often these are public and private developers, and in general proponents of projects, for the concept of applying environmental assessment to plans and policies is only slowly developing. The issue of externally financed projects clearly has important implications and the type of environmental assessment that should be undertaken by bilateral and multilateral aid agencies, and the conflicts that this may lead to with regard to national legislation, will also need to be considered;

-- Officials responsible for EIA. This group comprises many different actors who will require some form of training in environmental assessment. This includes managers and administrators who need to be made aware of the scope, utility and potential economic benefits of environmental assessment as well as those who may conduct all or part of the assessments. Given the complexity of producing such a study, where management and scientific skills have to be carefully combined in a multi-disciplinary team, the type of training required will vary greatly. The key project manager will require specialized management and environmental training, different from that of specialists, and where emphasis must be placed on how detailed studies such as air, noise, water, health, visual and ecological assessments can be integrated into the comprehensive environmental assessment;

-- Other control authorities. This group comprises officials in control authorities who, although not responsible for a country's environmental assessment system, may contribute to its effective operation. It will cover government officials responsible for pollution control, land use planning, conservation policies and scientists with special knowledge of detailed aspects of environmental assessment. The training of this group will be less intensive than for control authorities and project proponents.

-- Professional advisers on EIA and training institutions. This target group comprises several different parties. First, there are many consultant companies in developing countries who are now helping develop project proposals, and assisting control authorities in certain EIA tasks. Many of these are engineering and economic consulting companies diversifying into environmental assessment.

The second group comprises research institutes and academic institutions which are playing a greater role in the environmental assessment process by providing advice and technical support to either proponents or control/review authorities. Academic institutions also have a potentially key role to play in developing training programmes both at the undergraduate and post-graduate level and also in providing special training programmes. This target group can be described as one where the need is to "train the trainers" so that the concept of "sustainable training" can be developed. This should help to reduce the current reliance on expensive programmes in developed countries, but also reduce the number of "external" trainers "imported" from developed countries.

-- Environmental interest groups and the public. In developed countries the concept of public participation is now firmly established as an integral part of the environmental assessment process. Though this may at times be a contentious issue in developing countries, there is evidence to suggest that greater participation is now occurring and will increase in the future. It may therefore be necessary to provide some training for a range of groups that include: the business community; environmental groups and societies, including relevant NGOs; the media (television, radio and press); the public, including those who may be affected by project proposals; and schools (environmental studies generally).

Training requirements for the different target groups

The activities required to operate an environmental assessment process are broad and varied. The target groups identified will therefore have different training requirements. These are:

121

-- General awareness training. Training in this category will predominantly be for politicians, senior administrators, senior management in industry and business, and senior scientists. It should aim to provide a basic rationale for the utility of environmental assessment and an indication of the broad legal, procedural and methodological mechanisms. It should try to put over the basic concept that environmental assessment is not an activity designed to block economic growth, but a tool to ensure that economic benefits are maximised, adverse environmental impacts minimized and mitigated and that sustainable development is achieved. Training should be designed to ensure that decision makers and senior administrators are aware of the management processes required to achieve should environmental assessment: detailed knowledge of methods will not be required but an understanding of the usefulness of methods, as a scientific approach to project assessment, should be emphasized;

-- Training for environmental assessment project management. Whereas training of groups in the first category is designed to create an "environment of understanding" so that assessment can be undertaken in a specialized framework, training in this category is for those who will be responsible for the execution of environmental assessments. The form of this training will largely depend not only on existing knowledge but also previous experience. Therefore, for some groups, training will mean a refresher course on recent advances in the subject, while for others it will have to be basic and start from first principles. Within this category various types of training will be required:

-- Project managers. Management skills relating to project control, finance and co-ordination of specific technical inputs will be required. Though this type of person may be described as a generalist, he will require a comprehensive knowledge of environmental procedures and methods;

-- Technical experts. Given that environmental assessment normally requires that specialist types of assessment such as the prediction of noise, air or water impacts be undertaken, training for this group, which may comprise scientists from control authorities, consultants, project proponents, research institutes and academics, must attempt to show how specialized studies can be incorporated into an overall environmental assessment. Training therefore needs to focus on more technical aspects of environmental assessment with emphasis being placed on the "science" of EIA. Training will need to focus on methods (base line data and surveys, predictive and evaluation methods, monitoring and post audit studies) with less emphasis on legal and administrative aspects;

-- Review experts. Evidence from many developing countries indicates that one of the weakest links in the EIA process is project review. Whereas in a number of developed countries an independent review panel exists to advise, assist and review environmental assessments produced by a project proponent, it is normal in developing countries that the review be undertaken by a government agency. As a target group those who review environmental assessment are critical of the whole working of the process. Their training requirements call for both an overview of the EIA process and a detailed knowledge of assessment methods. The ability to know when to consult specialized experts on individual impact predictions and interpretations presented in an EIA will also need to be included in training programmes.

Types of training

Given the distinct target groups requiring various forms of training, it is now necessary to consider the types of training that exist, how they might be strengthened, whether new initiatives are necessary, and the form that these might take:

-- Undergraduate level training. There is little evidence to suggest that environmental assessment training currently exists at first degree levels in institutions of higher education in developing countries. In certain degree courses, such as engineering, environmental health, botany, geography, geology, etc., material is covered which is relevant as background knowledge for environmental assessment. This is also the case in a number of multi-disciplinary subjects such as environmental science and town planning.

It may not be appropriate to introduce separate first degree courses in environmental assessment given the benefits of initial training in a specific scientific subject. The introduction of the concept of EIA in those single and multi-disciplinary subjects should be encouraged to stimulate knowledge and potential interest in the concept.

-- Post-graduate training. It is at this level that positive steps could be taken by OECD Member states to encourage specific training in environmental assessment. In several countries, notably the United States, Canada, the Netherlands and Britain, environmental assessment is part of the training in land-use planning and environmental management degrees. This is also slowly taking place in a number of developing countries, in particular in the Far East.

It is recommended that two major initiatives be taken to develop environmental assessment training. First, a number of potential training centres should be identified in selected developing countries and special courses in environmental assessment be established, which could be supported with staff and finance by OECD Member states. Students could be selected from various relevant disciplines and a cadre of training experts formed who could develop the various management and technical aspects of environmental assessment.

Second, it is proposed that "awareness training" be introduced into specialised post-graduate courses which interface with environmental assessment. This could include science-based courses such as geology, soil science, pollution and technology control, ecology and health engineering, multi-disciplinary courses such as

123

land use and transportation planning, and social science courses such as economics and sociology;

-- Short course training. To alleviate the shortfall of experts in environmental assessment as soon as possible, it would appear that short courses geared to all target groups and covering both general awareness training and specialised components of the EA process, are a practical solution. Though in general these courses should operate at the national level, there is a case for more specialised courses being conducted on a regional basis.

General awareness training should be targetted at senior decision-makers and administrators. Given their seniority and availability, these courses should be of one or two days' duration. This target group could participate at the beginning and end of larger and more specialized courses so that they are familiar with the general context of the skills being developed by their specialist staff. The short courses for decision-makers should cover aspects of policy, management or environmental assessment and the economic benefits of environmental assessment.

Short courses for technical experts in the other target groups will need to take a variety of forms. The following topics would appear to be most appropriate but will need to be tailored to either the sectoral priorities of a given country or reflect distinct institutional procedures which may exist:

-- EIA procedures: geared to those who will operate the environmental assessment system with emphasis on legal and implementation considerations;

-- EIA methods: this type of course will emphasise the utility and application of environmental assessment methods and the integration of specific evaluative techniques (air, ecology, noise, pollution) into a structured environmental assessment system, relying heavily on the use of case studies to indicate the practical utility and application of environmental assessment methods;

-- Environmental assessment -- project management: the management of the EIA process. Skills to manage, finance, and communicate will be required.

Depending on the level of knowledge of environmental assessment in a developing country, and in particular where the concept is new, it may be necessary to provide training courses that combine all the three components listed above. After this it will be possible to develop more specialized courses tailored to particular sectors such as agriculture, forests, industrial development, and energy projects or specific facets of the EIA process. All of the above courses would require a period of study of from one to three weeks.

-- On-the-job training. The amount of on-the-job training in developing countries is currently very limited. It would seem useful to encourage this "hands on" training. One appropriate mechanism which OECD Member states could assist in developing would be to encourage proponents of projects, in particular bilateral and multilateral aid agencies and multinational corporations, to include a sytematic environmental training component as a contractual obligation. Wherever possible, those on short course training or post-graduate studies should be encouraged to spend a period of time in the project office of those conducting an environmental assessment to gain practical experience;

-- Training the trainers. As a priority action there would appear to be a pressing need to train certain members of selected key training institutions in developing countries on the multifarious aspects of environmental assessment. This should apply to those who will be encouraged to add aspects of environmental assessment to their training courses, normally at the post-graduate level, but in particular to those mounting specific EIA training programmes. A number of options are available:

-- Placement of trainers at established centres already teaching environmental management and assessment in developed countries;

-- Organising special training courses, probably on a regional basis, for course organisers and key contributors;

-- Short placements in agencies responsible for the environmental assessment process in both developed and developing countries and also in organisations conducting environmental assessment.

In all three options it would be necessary to tailor the training to the particular requirements of the country but the major objective must be to give the trainers practical rather than academic experience of environmental assessment. Indeed, one of the major criticisms of current training in EIA in Europe is that too theoretical an approach is adopted. This target group would benefit most from the specialized training courses, seminars and symposia held in developed countries.

Some Basic Principles for Training Programmes

It is neither practicable nor appropriate at this stage to define an "ideal" programme for the different components of environmental assessment training at various levels or for specific target groups. This is a special task which would require the bringing together of both EIA practitioners and trainers from developed and developing countries. A number of general principles that should be considered when formulating training initiatives, which can be summarised as follows:

-- It must be stated whether training is geared to generalists or specialists and this must be made explicit in the content of training programmes. Both may be desirable given that the institutional framework for EIA in countries varies greatly;

-- Wherever possible, training should be practical rather than theoretical. Field work and practical studies should be incorporated. If this is not possible, emphasis should be placed on case studies, practical exercises and simulation activities. Training must be geared towards implementation of sound EIA processes;

-- In general, the greatest training need is to develop the ability to undertake environmental assessments. Except where courses are specifically geared to legal/procedural requirements and environmental management, emphasis should be placed on both environmental assessment methodologies and the assessment of specific impacts;

-- Wherever possible, the trainers should have practical experience of conducting environmental assessments or be directly involved in the legal, procedural, or review aspects of the EIA process;

-- Ideally, the trainers should be nationals of the country where the training takes place. Initially, it may be necessary to use experts from other developing and developed countries who have practical experience both of EIA systems and training;

-- The length of short courses should be tailored to the specific needs of the target group. Awareness courses for senior personnel should probably be of one or two days duration. More technical courses would normally run from two to three weeks.

Training of Personnel from Developing Countries in Developed Countries

Currently, a limited number of personnel from developing countries are undertaking some form of training in environmental assessment in developed countries. The type of course varies greatly but includes:

-- Post-graduate or short courses geared to participants from the developed countries which can be attended by participants from developing countries;

-- Specific courses exclusively mounted for developed country participants. The majority of these courses only teach environmental assessment as a component and last from three to ten months;

-- Seminars and workshops that last from one week to three months on general or specific aspects of EIA, normally attended by participants from both developed and developing countries.

The question arises as to the potential benefits and drawbacks of training participants from developing countries in developed countries. According to evaluations made by those who have participated in such training, major benefits include:

-- Training relevant to the participants' needs which is not available in their own country;

-- An exchange of ideas with participants and faculty from both developed and developing countries;

-- An ability to concentrate on training and not be diverted by professional or domestic duties;

-- An opportunity to gain practical experience of environmental assessment in appropriate departments, agencies, research institutes and organisations. This is something which is increasingly encouraged by bilateral aid agencies that have sponsored participants in training courses;

-- Access to source materials and technical equipment relating to assessment which may not yet be available in the trainee's country.

There are also a number of drawbacks, which include:

-- The cost of training, and travel, can be very high;

-- The relevance of the training may be diminished given cultural and technical contrasts that may exist;

-- There is no guarantee that those who have access to funding for foreign training are necessarily the most appropriate persons.

Given the current shortage of trained personnel in environmental assesment it would appear that more training initiatives should be taken in both developed and developing countries. In EEC member states the Directive on Environmental Assessment, which went into effect in July 1988, is acting as a catalyst for a number of training initiatives. It is likely therefore that capacity for training personnel from developing countries will be increased.

To maximise this training potential, a number of actions should be taken:

-- Aid agencies and environmental ministries/departments should be encouraged to play a more active role within their own country in training personnel from developing countries, by mounting specific training modules and/or through financial support, provision of faculty and establishing strong working relationships with key environmental training centres in their own country. Training packages could be developed with personnel from aid agencies, environmental ministries and training institutes which could be presented in selected developing countries, or on a regional basis, and include faculty experts from developing countries concerned. This could be a relevant and cost-effective form of training and would have the added benefit of enabling aid agency and environment departmental officials to work closely together;

-- A proposal by a number of European environment training centres has recently been made to establish a Federation of European Environment Training Centres to encourage the exchange of personnel between countries, develop training aids for environmental assessment, to act as a catalyst in encouraging training in developing countries and to assist personnel coming to Europe from developing countries to receive the most appropriate training in the most appropriate institutions. Federations such as this could be encouraged by aid

127

agencies in OECD states to play a more important role in training institutions.

Training Aids

In addition to trained personnel, another priority requirement in developing countries is for environmental assessment training aids, including:

-- Books, information and data on environmental assessment. OECD Member states could provide valuable´ assistance in establishing environmental assessment information focal points in selected developing countries;

-- Evidence of "good practice" in environmental assessment. This could include EIA case studies and existing guidelines from developed countries. Certain countries such as the United States, through the Environmental Protection Agency, already donate much information to countries in the Third World. Information is required which is relevant to these countries, another initiative that OECD Member states could consider implementing;

-- Although a number of training manuals and training packages exist, there appears to be a need to develop this form of training aid. In this, aid agencies, environmental departments and training insti- tutes in developed countries could be encouraged to collaborate and produce both general and specific packages, including audio-visual material;

-- It has been shown from the evaluation of environmental assessment training programmes that one of the most successful training devices is simulation games. Some of the most successful games include a role-playing component. There is a need to develop simulations designed specifically for the needs of developing countries. Collaboration among aid agencies, environmental departments and trainers in developed and developing countries to produce relevant and appropriate examples would appear to be an important initiative.

PART IV

CONCLUSIONS

AN AGENDA FOR THE FUTURE

by
Joseph C. Wheeler
Chairman, OECD Development Assistance Committee

The environmental problems of developing countries are becoming increasingly severe:

-- World population is growing by over 80 million each year -- it is expected to increase from today's five billion to over 10 billion in the next century;

-- Resources are needed for food, shelter, and other requirements of a decent standard of living;

-- Pressure on land, water and forestry resources is increasing;

-- New technologies and products marketed internationally are demanding sophisticated environmental control strategies;

-- Most people are relying heavily on fuelwood and other renewable biomass sources of fuel;

-- Urbanisation is proceeding at explosive rates;

-- The management and analytical demands on governments and on the skilled manpower base are accelerating; and

-- Resources are exceptionally tight in the context of sluggish world growth, deteriorating terms of trade for developing-country exports and accumulated debt.

The consequences of these and other stresses on the natural environment and on the social and economic fabric of developing countries have been set out in the recent report "Our Common Future" of the World Commission on Environment and Development.

The deterioration of our natural environment at best represents a borrowing from future generations. At worst, if damage is irremediable, it robs us of our options.

This deterioration must be stopped, and development must become sustainable. The purpose of this Seminar was to deepen the dialogue among

OECD country and developing country environmental and development specialists to find answers to our common problems. OECD donors are determined to give environmental aspects of development increasing attention in their relations with developing countries. For their part, developing countries have increasingly come to the view that environmental issues must be addressed with urgency.

The following conclusions are linked to the three objectives of the Seminar.

Conclusions of the First Objective

The first objective of the Seminar was:

"To call upon developing country representatives to identify major environmental concerns and the policies, approaches and institutions which are needed to address them."

It was concluded that the environmental concerns vary considerably within and among developing countries and are often very different from those in OECD countries. Environment and development issues, therefore, must be viewed at the individual country or regional level before deciding on particular policies, approaches or institutional requirements.

Despite the unique characteristics of each individual developing country, environmental problems are, generally speaking, of two kinds:

-- First, and most important, is the destruction of the natural resource base due to deforestation, soil loss or degradation, desertification, water loss and salinisation;

-- Second, rapidly emerging problems are arising from industrialisation and urbanisation, including air and water pollution, lack of adequate sewerage and waste disposal systems and unhealthy living conditions often found in congested city slums.

Many, if not all, countries are now "borrowing" their environmental capital. Today's rates of natural resource consumption deny the next generation many of the environmental goods and services we currently enjoy.

This environmental deterioration is caused by a complex of interacting economic and social causes, many of which feed upon themselves to compound the cycle. Too many decisions are based on short-term factors leading to growing environmental stress which in the long run undermine the development process itself.

The affected people must participate in defining and understanding the problems and in planning and implementing the solutions. This requires a heavy emphasis on equity considerations in development and the need to decentralise decision-making processes. The example was used of an Indian village where the inhabitants were unable to control their environment, including reforestation and controlled grazing, until all parts of the village population, including the landless, were sharing the benefits.

People thus have to be given a stake in the benefits if they are to share in the solutions. This implies fundamental changes in land tenure and, more generally, the spreading of the benefits of development. The link between the quality of the environment and the development process must be made clear.

In identifying the policies, approaches and institutions which developing countries need in order to address their environmental concerns, it was concluded that the experience of OECD Member countries in integrating environmental policies with other economic and sectoral policies is relevant. This process of integration relies on the following factors:

-- "Internalisation", i.e. a better reflection of the environmental costs in the prices of resources used, based upon the application of the Polluter Pays Principle and the User Pays Principle and an elimination of the various governmental interventions which lead to under-priced environmental and natural resources;

-- A recognition that sound environmental planning can often save both government budgets and the public money in the short run -- for example, through fuel efficiency and less resource-intensive designs. Also, by incorporating environmental considerations in the project design from the beginning, the cost of the analysis can often be substantially reduced by data collection in economic and engineering studies and by avoiding decision-making delays;

-- The identification of complementary objectives between environmental policies and other economic and sectoral policies and an improved co-ordination between relevant authorities through more appropriate institutional arrangements;

-- The improvement of the information and analytical tools needed in decision-making;

-- Increasing public participation through education and awareness-building activities to ensure that the development process itself is based on local cultural, social and traditional values;

-- The cost advantages of preventing, rather than curing, environmental damage;

-- The full participation of the private sector, with market-oriented strategies, in all environment-relevant aspects of development projects and programmes.

It was concluded that OECD Member country aid agencies should further orient their development assistance policies and approaches towards provision of technology and systems appropriate to each recipient's circumstances and in the short run to put a heavy emphasis on maintenance and renewal. This must be accompanied by efforts which help foster sustainable development and full participation on the part of local populations. This means institution building, use of local capacity and continuing priorities for training.

Nevertheless, it was also recognised how difficult this process of change is. The integration of environmental and social concerns into the development assistance process demands, as spelled out above:

-- A multi-sectoral approach;

-- Planning across watersheds, agro-zones and ecological systems;

-- The adaptation of technologies to the capacity and needs of local populations;

-- Mechanisms that result in the increased involvement of the public;

-- Donor co-operation;

-- The internalisation of environmental costs so as to facilitate the recovery of maintenance and renewal costs.

The overriding need for developing countries to lead the process is clear. They must build environmental and natural resource management into their planning processes. In this context, the conclusions of the 1986 DAC High-Level meeting on "New Emphasis on Aid for Improved Development Polices and Programmes and on Strengthened Aid Co-operation" were found of particular relevance since they stress that "developing countries themselves are responsible for setting their policies and priorities and central responsibility for aid co-operation lies with each recipient government".

Conclusions on the Second Objective

The second objective of the Seminar was:

"To explore with developing countries how co-operation can be improved in the context of donor-assisted projects and programmes and to help them address the environmental problems identified."

In regard to this objective the participants at the Seminar looked particularly at the areas of pest management, water management and rural development and identified three ways in which co-operation can be improved:

Firstly, there are special projects and programmes designed to help restore, protect and improve the environmental basis for development in developing countries. Examples of such projects/programmes include:

-- Pest management involving the identification of the environmental consequences of existing inappropriate pest management approaches and techniques; the search for alternatives such as biological control methods; joint consideration of approaches for effective management, better screening, monitoring and control of pest management techniques; the development and enforcement of pesticide regulatory programmes, including adequate control of imports, packaging, labelling and effective enforcement; co-operation between donor agencies and developing countries in order to promote appropriate Integrated Pest Management methods; ensuring that economic incentives are fostering these improvements;

-- Watershed planning and management strategies might include ensuring that regulatory mechanisms are sufficient and enforced to guarantee balanced use and development of water resources; developing national or regional water plans (describing resources, defining demands, fixing user-priorities) which are regularly monitored and adjusted; creating autonomous and representative bodies responsible for water resources such as watershed areas; ensuring that in compensation systems, rights and duties, benefits, responsibilities and constraints are clearly established and evenly distributed.

-- In integrated rural and environmental management, strengthened international co-operation is needed to support environmental protection programmes and measures as well as integrated environment/development approaches including agro-ecological zoning, agro-forestry and forestry-pastoral systems.

Secondly, there is a need for a more integrated approach to development assistance. Examples of areas where progress can be made in this regard are initiatives to be taken by the developing countries to co-ordinate development assistance within their territory; better co-ordination among donors and within individual developing countries; incorporating environmental factors in the decision-making process; an evaluation of the impact of all projects and programmes which goes beyond a consideration of effects on the natural environment to include socio-economic and cultural aspects as well. Throughout, emphasis was put on the need for environmental/economic appraisals at all levels (i.e. project, programme, regional, national).

Thirdly, special programmes are needed for strengthening the institutional and professional capacities for environmental management in developing countries. Technical assistance can help in the preparation or revision of environmental legislation, environmental regulations and quality standards, and in the development of institutions needed for their implementation. Help can be provided to build the capacity to formulate and incorporate the environmental component into the central planning process. Environmental monitoring systems and mechanisms for ensuring that the information so gained is used in the decision-making process need to be established. Institutions are required to strengthen environmental and natural resources planning and management and help can be provided by financing research studies and programmes, organising training courses and generally building technical expertise on a long-term basis. Programmes for strengthening public awareness as part of the participatory process in development can be supported.

Conclusion on the Third Objective

The third objective of the Seminar was:

"To consider the most appropriate approaches for carrying out environmental assessments on donor-assisted projects and programmes."

In carrying out environmental assessments of development projects and programmes recipient governments and aid agences should:

-- Initiate assessments early in the project design process concurrently with feasibility studies; this will permit the environmental factors to influence design and save money by co-ordinated collection of data;

-- To the extent practical, actively involve officials of all interested parties in conducting the assessment (e.g. the use of joint teams comprised of personnel from the host country and donor agencies and use of joint venture consulting organisations);

-- Conduct "scoping" sessions prior to conducting assessments in order to define with the involvement of relevant ministries or agencies the essential elements of the assessment and the most cost-effective approach to carry it out;

-- Ensure that the assessment process continues beyond the point at which a decision to proceed with the project is taken, to include monitoring of the activity during its construction and operation.

In addition to the above conclusions relating to the three objectives of the Seminar, there was a consensus that OECD countries should continue to strengthen environmental co-operation with developing countries by:

-- Taking immediate steps to implement the OECD Council Recommendations on Environmental Assessment and Development Assistance;

-- Studying the ways in which developing countries can be aided in better identifying and incorporating environmental factors in economic decision-making;

-- Supporting the exchange of information among aid institutions and developing countries on the growing experience in dealing with environmental issues;

-- Promoting early actions to strengthen environmental co-operation between OECD Member countries and developing countries along the lines highlighted above.

Attention was also drawn to the impact on the environment of the severe current financial constraints on developing countries arising from deteriorating terms of trade and heavy debt burdens.

RECOMMENDATION OF THE COUNCIL

on environmental assessment of development assistance
projects and programmes

(adopted by the Council at its 627th Meeting
on 20th June 1985, C(85)104)

THE COUNCIL,

Having regard to Article 5 (b) of the Convention on the Organisation
for Economic Co-operation and Development of 14th December 1960;

Having regard to the Recommendation of the Council of 8th May 1979, on
the Assessment of Projects with Significant Impact on the Environment
[C(79)116];

Having regard to the Declaration on Anticipatory Environmental Policies
of 8th May 1979, adopted by the Environment Committee at Ministerial level;

Recalling in particular paragraphs 1 and 10 thereof, in which
Governments of OECD Member countries and Yugoslavia declared that "They will
strive to ensure that environmental considerations are incorporated at an
early stage of any decision in all economic and social sectors likely to have
significant environmental consequences" and that "They will continue to
co-operate to the greatest extent possible, ... will all countries, in
particular developing countries in order to assist in preventing environmental
deterioration";

Considering that many Member and non member countries have accumulated
over the years a growing body of experience in assessing environmental effects
of projects in their countries;

Mindful of the need for Member countries to adopt a common set of
principles when dealing with environmental issues and to bring support and
assistance to the use of environmental assessment in developing countries;

Recognising that, while developing countries have the responsibility
for managing their own environment, Member country aid agencies should, when
necessary, carry out environmental assessment and, in doing so, seek active
participation of the host Government;

On the proposal of the Environment Committee supported by the Development Assistance Committee;

I.   RECOMMENDS that Member Governments ensure that:

(a)   Development assistance projects and programmes which, because of their nature, size and/or location, could significantly affect the environment, should be assessed at as early a stage as possible and to an appropriate degree from an environmental standpoint;

(b)   When examining whether a specific development assistance project or programme should be subject to in-depth environmental assessment, Member country aid agencies should pay particular attention to those projects or programmes referred to in the Appendix, bearing in mind the particular legislative and socio-economic setting and environmental conditions in the host country;

(c)   Where dangerous substances or processes are involved, they also continue to seek ways to promote the integration of the best techniques of prevention and protection and the best manufacturing processes in projects in which they and their industrial enterprises are involved.

II.   INSTRUCTS the Environment Committee, in the light of practical experience of aid agencies in Member countries and in co-operation with the Development Assistance Committee, to prepare guidance on the types of procedures, processes, organisation and resources needed to facilitate the assessment of environmental effects of development assistance projects and programmes and to contribute to the early prevention and/or mitigation of potentially adverse environmental effects of certain aid projects or programmes.

(Appendix I)

PROJECTS AND PROGRAMMES MOST IN NEED OF ENVIRONMENTAL ASSESSMENT

Projects and programmes which are most in need of the environmental assessment can be identified on the basis of a number of criteria which aim at ascertaining whether the anticipated direct or indirect effects of a project or programme on the environment are likely to be significant.

When judging whether a specific project or programme may have a major effect on the environment, it is necessary to take into account, among other things, the ecological conditions in the area where it is planned to locate the project or programme. In-depth environmental assessment is always needed in certain very fragile environments (e.g., wetlands, mangrove swamps, coral reefs, tropical forests, semi-arid areas). When carrying out environmental assessment, issues which should be considered include effects on:

a. Soils and soil conservation (erosion, salination, etc.);

b. Areas subject to desertification;

c. Tropical forests and vegetation cover;

d. Water sources;

e. Habitats of value to protection and conservation and/or sustainable use of fish and wildlife resources;

f. Areas of unique interest (historical, archaelogical, cultural, aesthetic, scientific);

g. Areas of concentrations of population or industrial activities where further industrial development or urban expansion could create significant environmental problems (especially regarding air and water quality);

h. Areas of particular social interest to specific vulnerable population groups (e.g., nomadic people or other people with traditional lifestyles).

Projects or programmes most in need of environmental assessment fall under the following headings:

a. Substantial changes in renewable resource use (e.g., conversion of land to agricultural production, to forestry or to pasture land, rural development, timber production);

b. Substantial changes in farming and fishing practices (e.g., introduction of new crops, large scale mechanisation); use of chemicals in agriculture (e.g., pesticides, fertilizers);

c. Exploitation of hydraulic resources (e.g., dams, irrigation and drainage projects, water and basin management, water supply);

d. Infrastructure (e.g., roads, bridges, airports, harbours, transmission lines, pipelines, railways);

e. Industrial activities (e.g., metallurgical plants, wood processing plants, chemical plants, power plants, cement plants, refinery and petrochemical plants, agro-industries);

f. Extractive industries (e.g., mining, quarrying, extraction of p eat, oil and gas);

g. Waste management and disposal (e.g., sewerage systems and treatment plants, waste landfills, treatment plants for household waste and for hazardous waste).

The above list of projects or programmes is not in any order of importance and is not meant to imply that any particular project or programme type is necessarily more in need of environmental assessment than another. In addition, the list is not meant to be exhaustive as there may be projects or programmes not mentioned above which may still have significant effects on the environment in certain areas. Although the presence of a project or programme on the above list does not imply that such a project or programme will necessarily have significant adverse effects on the environment and some indeed have positive environmental effects, experience has shown that there is often a need to take particular measures to eliminate or mitigate the adverse environmental consequences of such projects or programmes. Whether a project or programme should be subject to in-depth environmental assessment will therefore depend on an analysis of all the facts of the specific case.

RECOMMENDATION OF THE COUNCIL

on measures required to facilitate the environmental assessment
of development assistance projects and programmes

(Adopted by the Council at its 649th Meeting
on 23rd October, 1986)

THE COUNCIL,

Having regard to Article 5b) of the Convention on the Organisation for Economic Co-operation and Development of 14th December 1960;

Having regard to the Recommendation of the Council of the 8th May 1979, on the Assessment of Projects with Significant Impact on the Environment [C(79)116];

Having regard to the Declaration on Anticipatory Environmental Policies of 8th May, 1979, adopted by the Governments of OECD Member countries and of Yugoslavia at a meeting of the Environment Committee at Ministerial level [C(79)121, Annex];

Having regard to the Declaration on "Environment: Resource for the Future" of 20th June 1985, adopted by the Governments of OECD Member countries and of Yugoslavia at a meeting of the Environment Committee at Ministerial level [C(85)111];

Recalling in particular paragraphs 1 and 11 of the latter Declaration, in which Governments of OECD Member countries and Yugoslavia declared that they will extend the use of environmental impact assessment and appropriate economic instruments, on the one hand, and strengthen their efforts to contribute to environmentally-sound development in developing countries, on the other hand;

Having regard to the Recommendation of the Council of 20th June 1985, on Environmental Assessment of Development Assistance Projects and Programmes [C(85)104];

Mindful of the need for Member countries to take into account the possible impacts of their activities on the environment and strive for closer cooperation with developing countries;

Recognising that environmental assessment of development assistance projects and programmes can help reduce the risk of costly and potentially adverse effects on the environment;

Recognising from the experience in Member countries that a successful environmental assessment process is dependent upon effective organisation, procedures and resources;

On the proposal of the Environment Committee and the Development Assistance Committee;

I.   RECOMMENDS that Governments of Member countries:

a)   Actively support the formal adoption of an environmental assessment policy for their development assistance activities;

b)   Examine the adequacy of their present procedures and practices with respect to implementing such a policy;

c)   Develop, in the light of that examination and to the extent necessary, effective procedures for an environmental assessment process taking into account, as need be, the approach outlined in Annex I;

d)   Firmly establish the responsibility for applying such procedures within each office responsible for the planning and/or implementation of development assistance projects and programmes;

e)   Establish the responsibility for supervising and providing guidance on the environmental assessment process in a central office of their aid agencies;

f)   Ensure that adequate human and financial resources are provided to conduct the environmental assessment process in a timely and cost-effective way; and

g)   Ensure the provision of human and financial resources to developing countries wishing to improve their capability for conducting environmental assessments, taking into account all or part of the measures outlined in Annex II.

II.   INVITES Member countries to exchange information on their progress in and experience with implementing environmental assessment on development assistance projects and programmes.

III.   INVITES the Development Assistance Committee in cooperation with the Environment Committee to:

a)   Collect further information on the way in which aid agencies of Member countries conduct environmental assessment of their development assistance projects and programmes;

b)   Examine how risk assessment can be incorporated in assessing the environmental effects of certain development assistance activities;

c)   Prepare a report in three years' time on all measures which will have been taken to implement this Recommendation and on pertinent activities in other international organisations.

IV.   INSTRUCTS the Secretary-General to transmit this Recommendation and its accompanying Report [ENV(85)27] to competent international organisations with a view toward fostering better environmental assessment of development assistance projects and programmes by all countries.

(Appendix I)

SUGGESTED APPROACH IN ESTABLISHING AN ENVIRONMENTAL
ASSESSMENT PROCESS FOR DEVELOPMENT ASSISTANCE ACTIVITIES

Whether a new process for assessing the environmental impacts of development assistance activities is created, or existing procedures are adapted to such a process, it is suggested that environmental assessment be coordinated with the host country government; integrated at an early stage of project and programme planning; reflected in the implementation of the activity and followed up by monitoring and post-audit evaluation.

The following elements of such a process have been found useful:

a)   An initial screening process should be undertaken to determine whether or not a full environmental assessment is required.

b)   An environmental assessment on a project or programme should begin at the pre-feasibility or project proposal stage and be integrated with cost-benefit and engineering feasibility studies.

c)   The content of the assessment should be determined by a procedure designed to identify reasonable project/programme alternatives and the most significant environmental impacts associated with them. The reason for doing so is to ensure that the ensuing assessment is carried out in the most timely and cost-effective manner by addressing only the most important issues necessary for making a decision. The procedure should be implemented preferably with a group of individuals responsible for the project or programme, coming together to discuss the issues and determine those to be addressed in the assessment. Host-government officials and, to the extent possible, the public affected by the activity and other interested parties should be included in the procedure as well.

d)   After this, terms of reference should be drawn up for the assessment itself. Depending on the size, nature and location of the project/programme, the assessment can range from a one to two page analysis based on existing information and carried out by a single individual to a comprehensive environmental impact statement based on extensive field surveys and data gathering and carried out by an interdisciplinary team. Regardless of the extent of the assessment, it is necessary that it be carried out in conjunction with traditional investigations (e.g. engineering feasibility).

e)   An assessment should not only point out the possible environmental consequences of a particular activity but also suggest mitigating (i.e. corrective) measures or alternative designs for limiting negative environmental impacts should the project/programme be implemented. In addition, attention should be given to the creation of appropriate institutional mechanisms in the host countries to ensure that mitigation measures are carried out.

f)   The assessment process should continue beyond the point at which a decision is taken, to include monitoring of the activity during its construction and operation. Monitoring is necessary to ensure that the findings of the assessment (e.g. suggested mitigating measures) are implemented, and to test the accuracy of the predictions made (e.g. the actual impact of the project on air quality, water quality, human health, ecosystem stability). The results of monitoring can lead to project modification as well as improving the data base for implementing the procedure described in paragraph (c) above in connection with future projects/programmes of a similar nature.

SUGGESTED MEASURES BY MEMBER COUNTRIES FOR IMPROVING THE CAPABILITY OF DEVELOPING COUNTRIES TO CONDUCT ENVIRONMENTAL ASSESSMENT

The ultimate goal of an aid agency environmental assessment process should be to help developing countries themselves manage their own development in an environmentally sound way. The following measures are suggested as steps which could be taken by aid agencies in Member countries in transfering to the developing world and supporting in it an environmental assessment capability.

An immediate measure which can be taken would be to involve actively host country officials in conducting environmental assessments for which aid agencies are responsible. That involvement could begin by including host government officials and others in the initial phase of the environmental assessment process and continue by engaging host country nationals in conducting the assessment and in monitoring activities (see Annex I).

OECD Member countries' aid and environmental agencies could institute training courses in environmental assessment. The provision of training should be made to a number of target groups in the host countries including elected representatives and senior decision-makers in government and business, high level administrators, project managers, technical specialists, members of review bodies and representatives of environmental interest groups. The specific type of training to be undertaken would vary depending on the target group. For policy makers, for example, seminars should be conducted to demonstrate the negative effects which result from a failure to incorporate environmental elements in economic development planning and emphasize the benefits to be gained from environmentally sound planning. Training for project managers and technical specialists would emphasize procedures and methods for environmental assessment and their role and significance in environmental management.

OECD Member countries might consider direct support to developing countries by providing environmental advisers to work with national planning agencies for as long as requested. Such advisers would have the task of helping government officials assess the environmental impacts that might be expected to arise from projects, programmes or policies and to inform decision-makers and the public of reasonable alternatives which would mitigate negative environmental impacts and enhance the quality of the human environment in the affected area.

146

The lack of adequate baseline data and information on the state of the environment is a major constraint to successfully implementing environmental assessment in developing countries. OECD Member countries' aid and environmental agencies might consider providing information such as host-country "environmental profiles" and base line studies on particularly sensitive areas. In addition, direct financial and technical assistance could be provided to host countries to carry out their own studies.

WHERE TO OBTAIN OECD PUBLICATIONS
OÙ OBTENIR LES PUBLICATIONS DE L'OCDE

ARGENTINA – ARGENTINE
Carlos Hirsch S.R.L.,
Galería Guemes, Florida 165, 4° Piso,
1333 Buenos Aires
Tel. 30.7122, 331.1787 y 331.2391
Telegram.: Hirsch-Baires

AUSTRALIA – AUSTRALIE
D.A. Book (Aust.) Pty. Ltd.
11-13 Station Street (P.O. Box 163)
Mitcham, Vic. 3132 Tel. (03) 873 4411
Telex: AA37911 DA BOOK Telefax: (03)873.5679

AUSTRIA – AUTRICHE
OECD Publications and Information Centre,
4 Simrockstrasse,
5300 Bonn (Germany) Tel. (0228) 21.60.45
Telex: 8 86300 Bonn Telefax: (0228)26.11.04
Gerold & Co., Graben 31, Wien 1 Tel. (1)533.50.14

BELGIUM – BELGIQUE
Jean de Lannoy, Avenue du Roi 202
B-1060 Bruxelles Tel. (02) 538.51.69/538.08.41
Telex: 63220

CANADA
Renouf Publishing Company Ltd
1294 Algoma Road, Ottawa, Ont. K1B 3W8
Tel: (613) 741-4333
Telex: 053-4783 Telefax: (613)741.5439
Stores:
61 Sparks St., Ottawa, Ont. K1P 5R1
Tel: (613) 238-8985
211 rue Yonge St., Toronto, Ont. M5B 1M4
Tel: (416) 363-3171
Federal Publications Inc.,
165 University Avenue,
Toronto, ON M5H 3B9 Tel. (416)581-1552
Telefax: (416)581.1743
Les Publications Fédérales
1185 rue de l'Université
Montréal, PQ H3B 1R7 Tel.(514)954.1633
Les Éditions la Liberté Inc.,
3020 Chemin Sainte-Foy,
Sainte-Foy, P.Q. G1X 3V6, Tel. (418)658-3763
Telefax: (418)658.3763

DENMARK – DANEMARK
Munksgaard Export and Subscription Service
35, Nørre Søgade, P.O. Box 212148
DK-1016 København K Tel. (45 1)12.85.70
Telex: 19431 MUNKS DK Telefax: (45 1)12.93.87

FINLAND – FINLANDE
Akateeminen Kirjakauppa,
Keskuskatu 1, P.O. Box 128
00100 Helsinki Tel. (358 0)12141
Telex: 125080 Telefax: (358 0)121.4441

FRANCE
OCDE/OECD
Mail Orders/Commandes par correspondance :
2, rue André-Pascal,
75775 Paris Cedex 16 Tel. (1) 45.24.82.00
Bookshop/Librairie : 33, rue Octave-Feuillet
75016 Paris
Tel. (1) 45.24.81.67 or/ou (1) 45.24.81.81
Telex: 620 160 OCDE Telefax: (33-1)45.24.85.00
Librairie de l'Université,
12a, rue Nazareth,
13602 Aix-en-Provence Tel. 42.26.18.08

GERMANY – ALLEMAGNE
OECD Publications and Information Centre,
4 Simrockstrasse,
5300 Bonn Tel. (0228) 21.60.45
Telex: 8 86300 Bonn Telefax: (0228)26.11.04

GREECE – GRÈCE
Librairie Kauffmann,
28, rue du Stade, 105 64 Athens Tel. 322.21.60
Telex: 218187 LIKA Gr

HONG KONG
Government Information Services,
Publications (Sales) Office,
Information Services Department
No. 1, Battery Path, Central
Tel.(5)23.31.91 Telex: 802.61190

ICELAND – ISLANDE
Mál Mog Menning
Laugavegi 18, Pósthólf 392
121 Reykjavik Tel. 15199/24240

INDIA – INDE
Oxford Book and Stationery Co.,
Scindia House,
New Delhi 110001 Tel. 331.5896/5308
Telex: 31 61990 AM IN Telefax: (11) 332.5993
17 Park St., Calcutta 700016 Tel. 240832

INDONESIA – INDONÉSIE
Pdii-Lipi, P.O. Box 3065/JKT.
Jakarta Tel. 583467
Telex: 73 45875

IRELAND – IRLANDE
TDC Publishers - Library Suppliers,
12 North Frederick Street,
Dublin 1 Tel. 744835-749677
Telex: 33530TDCP EI Telefax: 748416

ITALY – ITALIE
Libreria Commissionaria Sansoni,
Via Benedetto Fortini 120/10,
Casella Post. 552
50125 Firenze Tel. (055)645415
Telex: 570466 Telefax: (39.55)641257
Via Bartolini 29, 20155 Milano Tel. 365083
La diffusione delle pubblicazioni OCSE viene assicurata
dalle principali librerie ed anche da :
Editrice e Libreria Herder,
Piazza Montecitorio 120, 00186 Roma
Tel. 6794628 Telex: NATEL I 621427
Libreria Hœpli,
Via Hœpli 5, 20121 Milano Tel. 865446
Telex:31.33.95 Telefax: (39.2)805.2886
Libreria Scientifica
Dott. Lucio de Biasio "Aeiou"
Via Meravigli 16, 20123 Milano Tel. 807679
Telefax: 800175

JAPAN – JAPON
OECD Publications and Information Centre,
Landic Akasaka Building, 2-3-4 Akasaka,
Minato-ku, Tokyo 107 Tel. 586.2016
Telefax: (81.3) 584.7929

KOREA – CORÉE
Kyobo Book Centre Co. Ltd.
P.O.Box 1658, Kwang Hwa Moon
Seoul Tel. (REP) 730.78.91
Telefax: 735.0030

MALAYSIA/SINGAPORE – MALAISIE/SINGAPOUR
University of Malaya Co-operative Bookshop Ltd.,
P.O. Box 1127, Jalan Pantai Baru 59100
Kuala Lumpur, Malaysia/Malaisie
Tel. 756.5000/756.5425 Telefax: 757.3661
Information Publications Pte Ltd
Pei-Fu Industrial Building,
24 New Industrial Road No. 02-06
Singapore/Singapour 1953 Tel. 283.1786/283.1798
Telefax: 284.8875

NETHERLANDS – PAYS-BAS
SDU Uitgeverij
Christoffel Plantijnstraat 2
Postbus 20014
2500 EA's-Gravenhage Tel. (070)78.99.11
Voor bestellingen: Tel. (070)78.98.80
Telex: 32486 stdru Telefax: (070)47.63.51

NEW ZEALAND – NOUVELLE-ZÉLANDE
Government Printing Office Bookshops:
Auckland: Retail Bookshop, 25 Rutland Street,
Mail Orders, 85 Beach Road
Private Bag C.P.O.
Hamilton: Retail: Ward Street,
Mail Orders, P.O. Box 857
Wellington: Retail, Mulgrave Street, (Head Office)
Telex: COVPRNT NZ 31370 Telefax: (04)734943
Cubacade World Trade Centre,
Mail Orders, Private Bag
Christchurch: Retail, 159 Hereford Street,
Mail Orders, Private Bag
Dunedin: Retail, Princes Street,
Mail Orders, P.O. Box 1104

NORWAY – NORVÈGE
Narvesen Info Center – NIC,
Bertrand Narvesens vei 2,
P.O.B. 6125 Etterstad, 0602 Oslo 6
Tel. (02)67.83.10/(02)68.40.20
Telex: 79668 NIC N Telefax: (47 2)68.53.47

PAKISTAN
Mirza Book Agency
65 Shahrah Quaid-E-Azam, Lahore 3 Tel. 66839
Telegram: "Knowledge"

PORTUGAL
Livraria Portugal, Rua do Carmo 70-74,
1117 Lisboa Codex Tel. 347.49.82/3/4/5

SINGAPORE/MALAYSIA – SINGAPOUR/MALAISIE
See "Malaysia/Singapore". Voir «Malaisie/Singapour»

SPAIN – ESPAGNE
Mundi-Prensa Libros, S.A.,
Castelló 37, Apartado 1223,
Madrid-28001 Tel. 431.33.99
Telex: 49370 MPLI Telefax: 275.39.98
Libreria Bosch, Ronda Universidad 11,
Barcelona 7 Tel. 317.53.08/317.53.58

SWEDEN – SUÈDE
Fritzes Fackboksföretaget
Box 16356, S 103 27 STH,
Regeringsgatan 12,
DS Stockholm Tel. (08)23.89.00
Telex: 12387 Telefax: (08)20.50.21
Subscription Agency/Abonnements:
Wennergren-Williams AB,
Box 30004, S104 25 Stockholm Tel. (08)54.12.00
Telex: 19937 Telefax: (08)50.82.86

SWITZERLAND – SUISSE
OECD Publications and Information Centre,
4 Simrockstrasse,
5300 Bonn (Germany) Tel. (0228) 21.60.45
Telex: 8 86300 Bonn Telefax: (0228)26.11.04
Librairie Payot,
6 rue Grenus, 1211 Genève 11 Tel. (022)731.89.50
Telex: 28356
Maditec S.A.
Ch. des Palettes 4
1020 – Renens/Lausanne Tel. (021)635.08.65
Telefax: (021)635.07.80
United Nations Bookshop/Librairie des Nations-Unies
Palais des Nations, 1211 – Geneva 10
Tel. (022)734.60.11 (ext. 48.72)
Telex: 289696 (Attn: Sales) Telefax: (022)733.98.79

TAIWAN – FORMOSE
Good Faith Worldwide Int'l Co., Ltd.
9th floor, No. 118, Sec.2, Chung Hsiao E. Road
Taipei Tel. 391.7396/391.7397
Telefax: 394.9176

THAILAND – THAILANDE
Suksit Siam Co., Ltd., 1715 Rama IV Rd.,
Samyam, Bangkok 5 Tel. 2511630

TURKEY – TURQUIE
Kültur Yayinlari Is-Türk Ltd. Sti.
Atatürk Bulvari No. 191/Kat. 21
Kavaklidere/Ankara Tel. 25.07.60
Dolmabahce Cad. No. 29
Besiktas/Istanbul Tel. 160.71.88
Telex: 43482B

UNITED KINGDOM – ROYAUME-UNI
H.M. Stationery Office (01)873-8483
Postal orders only:
P.O.B. 276, London SW8 5DT
Telephone orders: (01) 873-9090, or
Personal callers:
49 High Holborn, London WC1V 6HB
Telex:297138 Telefax: 873.8463
Branches at: Belfast, Birmingham, Bristol, Edinburgh,
Manchester

UNITED STATES – ÉTATS-UNIS
OECD Publications and Information Centre,
2001 L Street, N.W., Suite 700,
Washington, D.C. 20036-4095 Tel. (202)785.6323
Telex:440245 WASHINGTON D.C.
Telefax: (202)785.0350

VENEZUELA
Libreria del Este,
Avda F. Miranda 52, Aptdo. 60337,
Edificio Galipan, Caracas 106
Tel. 951.1705/951.2307/951.1297
Telegram: Libreste Caracas

YUGOSLAVIA – YOUGOSLAVIE
Jugoslovenska Knjiga, Knez Mihajlova 2,
P.O.B. 36, Beograd Tel. 621.992
Telex: 12466 jk bgd

Orders and inquiries from countries where Distributors
have not yet been appointed should be sent to: OECD,
Publications Service, 2, rue André-Pascal, 75775 PARIS
CEDEX 16.

Les commandes provenant de pays où l'OCDE n'a pas
encore désigné de distributeur devraient être adressées à :
OCDE, Service des Publications. 2, rue André-Pascal,
75775 PARIS CEDEX 16.

72547-6-1989

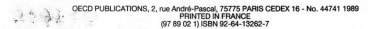
OECD PUBLICATIONS, 2, rue André-Pascal, 75775 PARIS CEDEX 16 - No. 44741 1989
PRINTED IN FRANCE
(97 89 02 1) ISBN 92-64-13262-7